Mines of the Eagle Country: Nickel Plate & Mascot

Mines of the Eagle Country: Nickel Plate & Mascot

Compiled and written
by Doug Cox

Word Processing & Design
by Ruth Burke

Cover Design:
Stuart Bish. M.P.A. Photography/Design

Printed by: Webco West. Penticton. B.C.
Cover printed by: Ehmann Printing Ltd.. Kelowna. B.C.

Copyright ©1997
Skookum Publications.
1275 Riddle Rd.. Penticton. B.C. V2A-6J6
Phone/Fax (250)492-3228
ISBN 0-919773-11-7

Mines of the Eagle Country:
Nickel Plate & Mascot

Canadian Cataloguing in Publication Data

Cox, Doug
  Mines of the Eagle Country; Nickel Plate & Mascot

Includes bibliographical references and index
ISBN 0-919773-11-7

  1. Nickel Plate Mine (B.C.) - History.
  2. Mascot Mine (B.C.) - History
  3. Gold mines and mining - Social aspects
     - Similkameen River Valley - Okanagan Valley
  4. Similkameen River Valley, Okanagan Valley - History
  1. Title,

  HD9536.C23S54 1997   971.1'5   C97-900290-7

Photo cover:
Restoration of the **Hedley Mascot Mine** *buildings overlooking Hedley.*
Greyback Construction Ltd. *of Penticton is the general contractor with*
helicopter service supplied by **Eclipse Helicopters Ltd.,** *also of Penticton.*
Photo by Stuart Bish, M.P.A. Photography/Design

Other books by Doug Cox...

OKANAGAN ROOTS

An historical look at the South Okanagan and Similkameen. This book, which contains a wealth of historical photos, covers the geography, legends, early settlement, mining, cattle raising and ranching, logging and orcharding. The book describes most settlements in the Similkameen Valley.
1987 ISBN 0-919773-04-4

RODEO ROOTS/50 Years of Rodeo in the Similkameen and Okanagan

This publication was produced to commemorate the 50th Anniversary of the Keremeos Elks Rodeo. The book deals with early cattlemen and horsemen, cattle drives and cattle sales, to present day rodeos. Included are the Oliver Stampede, Twin Lakes Stampede, Stelkia Rodeo, Chopaka Rodeo, All Indian Rodeo and Summerland Rodeo.
1988 ISBN 0-919773-05-2

WAGON TRAIN OVER THE MONASHEE

As a youth raised during Canada's depression years of the 1930's, Edward Patterson travelled with his family across Southern Alberta and B. C. in a cavalcade of three covered wagons and a team hitched to a 1928 Chevrolet truck, all followed by 250 head of unbroken range horses - the foundation stock for a new start in B.C.
1988 ISBN 0-919773-06-0

HERITAGE TOURS

This publication is a guide which will direct the history enthusiast on 6 tours designed to explore the history and heritage of the South Okanagan and Similkameen, with illustrated maps of each excursion. Each location is identified with an original historical photo and a description of the site. Suitable for armchair or auto tours.
1991 ISBN 0-919773-07-9

THE BEST OF THE WESTERN

Best of the last 10 year collection of historical photos featured weekly in the Western News Advertiser, containing 180 photos and significant captions relating to Penticton, the South Okanagan, Similkameen and Tulameen. The book covers early roads, lake boats, railways, ranching, logging, mining and the fruit industry.
1992 ISBN 0-919773-09-5

PIONEERING THE PEACE

In 1927, Frank Hunter, a youth from Manitoba, ventured northwest to the Peace River country of British Columbia - then one of the last frontiers of Canada. This is an account of his lifelong adventure as he surveyed for the Northern Alberta R.R., hauled freight for the Alaska Highway and wrestled a farm from the unforgiving north.
1993 ISBN 0-919773-08-7

S.S. SICAMOUS, Queen of Okanagan Lake

On British Columbia's inland lakes and rivers the vessel of choice was the steam powered, paddle wheel driven, "sternwheeler". The S.S. Sicamous was the finest, and last of these magnificent boats to be built. This is the story of the "Queen of Okanagan Lake" and the wharves of call on her daily tour of Okanagan Lake.
1995 ISBN 0-010773-10-9

Skookum Publications
1275 Riddle Rd., Site 176, Comp. 4
Penticton, B.C. V2A 6J6
Phone/Fax (250)492-3228

PREFACE

The Nickel Plate Mine complex was, for close to a century, the economic hub of the Similkameen Valley in Southern British Columbia. The mine, which was discovered in 1898, obtained all supplies by sternwheeler coming down Okanagan Lake. At Penticton, then a transfer point for freight, the supplies and machinery for the mine, were transferred to freight wagons where four and six horse teams hauled their loads the forty miles and the five thousand feet in elevation up to the mine.

Nickel Plate Mine and the townsite of Nickel Plate, located at the 6000 foot level in elevation, operated in a climate now reserved for ski resorts. To add to the snow and cold, most families lived in tent houses, constructed of a base of logs or lumber, with a tent roof. This housing arrangement was because of a mining regulation at the time. Homes at Nickel Plate had a supply of electricity generated from a power dam, flume, and turbine on the Similkameen River. Moderate comforts were enjoyed until the flume to the turbine froze during the winter, at which time the residents of the townsite were given twenty-four hours to evacuate. Spring thaw on the Similkameen resumed the flow of water through the flume, providing electricity, and the mine re-opened for another season.

The single men, who lived in the bunkhouses in the Nickel Plate townsite, commuted to the bright lights of Hedley, seven hotels at one time, by riding a piece of board with cleats attached to form a guide as they slid down the tram line rails to town. Gum boots against the rail acted as brakes! The trip home was an unauthorized ride up the ore skip, much faster than the forty-seven miles by automobile over the Old Nickel Plate Road.

Compiling a history of mining, such as Mines of the Eagle Country, is a compromise. There is a compromise between the amount of technical information that should be included, and the reminiscences of those who lived and worked at the mine. Hopefully, there is sufficient information to satisfy those who are miners, and want to read about the technical workings of a mine. There is, however, a social aspect, a feeling of camaraderie, of those who lived and worked at the mountaintop village and mine, often in less than ideal conditions, which is also the essence of this mine. There are certain inconsistencies in the tales told by the old timers, after all they were not metallurgists or mining engineers. They do, at times, even contradict each other! This is, however, how they remember their time at the Nickel Plate.

The Nickel Plate Mine was one of the oldest, longest producing gold mines in British Columbia. The Hedley Mascot Mine, high on the cliffs above Hedley, was one of our richest gold producing mines. This is the story of the people who lived and worked there.

Acknowledgements

This history of a mine, which has been in operation for close to one hundred years, would not have been possible without the assistance of many people. The staff at the Nickel Plate Mine were most helpful. Glenn McDonald, public relations officer, could always arrange a radio-equipped, 4WD vehicle, with a driver, whenever I requested a visit for a photo session. I would even phone to check that it was a bright sunny day before I came up to the mine. Jerry Leblanc, mine foreman, who often drove the truck, would wait patiently as I searched for the best location to take a photograph, and then set up a camera and tripod. One time the mine crew postponed their lunch hour so I could capture an ore truck emptying its load into the crusher. Glenn also patiently explained the workings of the mine and mill to this non-miner. Merlyn Royea, manager of the mine, shared his mining expertise and experience by editing the manuscript, and clarifying certain mining terms and mining techniques. Barry Given, metallurgist, and Brenda Dickson, environmental co-ordinator, discussed the technical aspects of the reclamation of the mine, before it is returned to its natural state, prior to one hundred years of mining

Betty Bork, writer and co-author on some historical articles and other publications, shared her expertise in the initial organization of the book. Brian Wilson, photo technicion, restored and re-photographed the heritage photos used in this publication. Norma Lippa, English teacher and friend, helped make the book reader-friendly. She added punctuation and sentence structure to make the taped and transcribed conversations of the former residents and employees more vibrant and informative. My wife, Joyce, gave invaluable assistance and support throughout this project. She was the final proofreader.

There were also the people who lived at the "top", Nickel Plate Townsite, worked in the mines, lived in Hedley, or worked at the Reduction Mill. Without these people sharing their experiences and recollections of their time at this mining complex in the Similkameen Valley, this book would not be.

Thank you all.

Doug Cox

TABLE OF CONTENTS

1. THE SIMILKAMEEN

In the Indian language is the word Similkameen, meaning "a large winding stream". Tumbling down the box canyons of the Ashnola Mountains to the south and the escarpment of the interior plateau to the north, are several tributaries which merge to flow through a steep walled valley or canyon, less than four miles across in some places. The Similkameen flows steadily south-eastward through beautiful Nighthawk Pass into the United States where it joins the Okanagan River at Oroville in the State of Washington. The climate of the Similkameen Valley is ideal. Free from extremes of temperature, yet having ample snowfall to provide moisture for vegetation, there is abundant water for a wide range of animals and trout-filled mountain lakes.

A mixture of Okanagan and Thompson tribes, the Interior Salish Indians first inhabited this land abundant with game, edible berries, roots, wild vegetables, and sparkling creeks teeming with trout. Throughout the summer season, the Indians lived in teepees, moving to the berry-picking areas and gathering roots and wild herbs. In late summer they carefully placed their weirs in the river for the salmon which moved upstream. Salmon dried in the hot sun of late summer, and preserved with wood smoke, was part of their winter diet.

The Similkameen Valley

As winter approached the Indians returned to their villages where they lived in small "keekwillie" houses, a semi-underground conical shaped structure made from saplings, bark, woven reeds, moss and dirt.

The Indians of the Similkameen were sought out by the tribes in the Okanagan and those farther north in the Shuswap, for two commodities which they were able to obtain. Red ocher, the basic writing material used to record everyday and historic events on rock faces was obtained from where the Tulameen River joins the Similkameen River. Princeton, the town which grew at the confluence of these two rivers, was originally named Vermillion Flats in recognition of the red earth available at that location.

The second commodity sought after by other tribes was eagle feathers, which were also used to distinguish an important event or deed. Eagle feathers, worn in the hair, signified one's rise to manhood, marital status, or one's heroic deeds. These feathers were obtained from the noblest of birds, which inhabited the rocky crags and outcrops high above the valley floor. This was the land referred to by the Indian people as the Eagle Country.

The 49th parallel between Canada and the United States of America was established in 1846 by the Oregon Treaty. It was a significant factor in the development of the Similkameen Valley, for the route of the Hudson's Bay Company had to be altered to accommodate the new post at Fort Langley, from the previously travelled Okanagan-Columbia River trail. This necessity eventually brought about the establishment of a Hudson's Bay Post in the Similkameen Valley, with Francois Deschiquette from Fort Okanagan in charge of locating the spot.

A location was chosen near what is now Cawston, and a station established by Chief Factor Roderick Finlayson in 1862. He was assisted very little by Deschiquette, as the man was ill and died at the Similkameen station in September of 1862. However, the trading of the lower Okanagan and Similkameen Indians had been secured with the presence of this post. A rough but practical wagon road existed between Princeton and the Okanagan Valley known as the Dewdney Trail, over which all freighting was done, but there were no towns located on this route.

In 1864 the Post was moved to Keremeos during the term of Chief Factor Roderick McLean. John Tait succeeded McLean in 1867, however, in 1872 the Similkameen station was closed forever.

In 1860, along Twenty Mile Creek, gold had been discovered and worked in small quantities, but not until 1898 was the full potential of the area realized.

With the abandonment of the Hudson's Bay Post, the fur trading era drew to a close. A new decade of industry in mining would, in a short period of time, expand to cattle ranching, logging, and the cultivation of vast acres of land for orchards and ground crops. It would bring about the birth and growth of three towns in the heart of the Similkameen Valley, with an elaborate freighting system to serve them.

It was as though the Similkameen Valley had waited patiently through the decades of slow development prior to 1900, for the discovery of gold in its rich mountain seams to happen. And, when it did, the heyday of Hedley, Nickel Plate, and Keremeos had begun!

2. Nickel Plate Discovered

At one point in the Similkameen Valley a creek called Twenty Mile Creek, tumbles down a boulder-strewn channel between Stemwinder and Nickel Plate mountain. Early travellers, particularly prospectors travelling over the Dewdney trail, gave the rock outcrops above the river, the folded stratified rock on Stemwinder, and the iron staining on Red Mountain close attention. All of these factors could indicate an ore deposit.

It is assumed there was an old trail that ran over the Nickel Plate Mountain, a shortcut first used by the fur men and voyageurs of the Hudson Bay Company. The trail was later used by miners gaining access to Camp McKinney, an early mining camp near what is now Oliver. In all likelihood that area had been prospected as the miners passed through, or even earlier, during the short placer mining rush of 1859 and 1860.

Edgar Dewdney, the roadbuilder, who later became the Honorable Edgar Dewdney, Lieutenant Governor, had James Riordan and Charlie Allison stake claims on what is now Nickel Plate mountain in 1894. Jimmy Riordan was a rancher at Olalla at the time. He claimed to have gained a knowledge of the area from Pinto, an enterprising Indian who lived on a river flat east of Hedley, now called Pinto Flats. Mt. Riordan is the name given to one of the mountains of the Apex Alpine ski area. J.O. Coulthard, better known as Ozzie, also had

Nickel Plate Mountain occupies the area to the east of the Twenty Mile Creek junction and as a consequence, is ringed by the cliffs of Twenty Mile Canyon on the west, and the Similkameen canyon on the south. The mountain slopes gently to the hanging valley of Eighteen Mile Creek on the east, while to the north it merges into the moderately hilly surface of the Interior Plateau. Hedley lies 1,700 feet above sea level; 4,500 feet above the town towers the summit of Nickel Plate. The height of the plateau is about 6,000 feet.

Herb Clark Estate Photo

The site of Hedley with Twenty Mile Creek in the foreground and the Similkameen River in the background. At Hedley, Twenty Mile Creek, a tributary, enters from the north front canyon, from 2,500 to 4,000 feet deep, with sides, gashed by many deep and narrow box canyons, sloping downward at steep angles. The Similkameen canyon is 4,000 feet in depth, and is frequently less then four miles from rim to rim.   Herb Clark Estate Photo

a claim on the mountain. However, all four claims were allowed to lapse; it was thought they were not worth proving!

Ozzie's reason for letting his claim lapse is rather unique. This occurred when apparently R.L. Cawston, for whom Cawston is named, and his brother were moving cattle to Rossland on their fall cattle drive. Since Rossland had an assay office, and R.L. and Ozzie were fellow cattlemen, this was an opportunity for Ozzie Coulthard to have his ore from the Nickel Plate Mountain assessed. When R.L. and his brother George reached the Columbia River, there was the usual problem of getting the steers into the river to swim across; to add to the confusion a paddlewheeler rounded a bend and sounded a warning whistle. When he had exhausted all methods known to him to try turning his panicked steers back, R.L. discovered some rocks in his pocket and threw them at the frightened steers. Somewhere on the shores of the Columbia River there are possibly some very high grade ore samples, if someone cares to take a look for them!

It was a few years later that two prospectors, Frances H. Wollaston and Constantine Arundel climbed to the top of Nickel Plate, and discovered the "free gold" in a rusty red outcrop of what is now the Nickel Plate mine.

Wollaston had been an assayer in Nevada. Arundel, another Englishman, had worked at the Bullock-Webster Ranch in Keremeos, and had done some prospecting. An earlier burn had cleared the timber and allowed the soil to erode, exposing a weathered outcrop of gossan. They tried panning and found gold "colors".

Pinto Flats, an area to the south between Hedley and the river, was named after Pinto, an Indian who owned the cabin. In the autumn, native folk met here to arrange hunting parties and smoke some of the meat for winter provisions.   Herb Clark Estate photo

The burn or forest fire that had occurred could have been as a result of earlier prospect work in the area. One tactic of the early prospectors was to light a fire on a mountain top, or in semi-controlled conditions, with the idea of exposing rock outcrops, or generally making the search for minerals easier. This is, of course, speculation; the fire could also have occurred from natural causes.

Wollaston and Arundel staked the Nickel Plate, Bulldog, Horsefly, Sunnyside, and Copperfield claims. A year earlier in 1897, Peter Scott located the Rollo claim. That year also, C. Johnson and A. Jacobson grubstaked by the manager of Granby Mines, W.Y. Williams, staked the Mound and Copper Cleft claims. Peter Scott returned the next year and staked the Princeton, Warhorse, Kingston and others.

The results of this discovery must have been very thrilling to the prospectors. It would have been an even greater thrill had the two known that this was the beginning of one of the richest gold mines in Canada and it would produce gold for over fifty years, making Hedley the economic hub of the Similkameen! The discovery was also going to create an economic boom in Penticton, which was to be significant in the growth of that community.

In the autumn of 1898, Wollaston and Arundel displayed some of the surface ore samples at the coast. A popular version of this story has the men displaying their ore at the New Westminster Fair. George Cahill, a contemporary of the men, relates that Wollaston and Arundel sent samples to John R. Tool, manager of Marcus Daly interests in Montana. Tool wired M.K. Rodgers, who was waiting in Victoria to catch a boat to the Cassiar district. The meeting was no accident, the men were prospectors, and one had worked as an assayer; they

The staff at the Nickel Plate mine in those early years of development consisted of M.K. Rodgers, general manager; Wesley P. Rodgers, a brother of M.K. Rodgers, mine engineer and surveyor; Gomer P. Jones, who came to the Nickel Plate in August 1900, mine superintendent; and Frank Bragg, storekeeper and timekeeper..   Pat Wright photo collection

knew they had a good thing and they were looking for a buyer. Rogers cancelled his sailing to Stewart, and journeyed to the Similkameen to look at the claim first hand. These observations and the assays proved so satisfactory that he took a bond on the claims owned by Wollaston and Arundel for a reported $60,000.

The following story, taken from "Mother Earth's Treasure Vaults," issue of 1905, written by Percy F. Godenrath, will, perhaps, provide a clearer picture of the story of this really great mine:

"Hardships! You get used to hardships after travelling for three and one-half years, covering over 135,000 miles and sampling in the neighborhood of 700 mineral claims, from the tropical climes of Guatemala to the inhospitable shores of Alaska," laughingly retorted M.K. Rodgers, manager of the famous Nickel Plate mine in the Similkameen, in reply to my query for information regarding the life of a mining engineer in the hills.

"And disappointment? Yes, mining men have their full share, for Fortune is ever fickle to one who seeks to unlock her treasure chests," he resumed. "One instance in point will suffice before I tell you how I ran across the Nickel Plate. I started out from Butte in 1895, holding a sort of roving commission on behalf of the late Marcus Daly. I first tackled the then little known Boundary district, examining the many remarkable low grade ore bodies in Phoenix and Deadwood camps.

I had just made a flying trip to Australia and was in Victoria bound for the Skeena River country. Time hung heavily on my hands while waiting for the northbound steamer. One day

This was the cabin on the Camp Rest Trail, photographed in 1919, one of the first trails which led to the Nickel Plate mine. It was on this trail that George Cahill led a 35-horse pack train from Fairview to the Nickel Plate mine for M.K. Rodgers in 1898. The rider is Billy Daly of Keremeos.

Estabrooks photo

I happened to be in William Wilson's store, on Government street. He showed me some striking-looking samples of ore that he said came from the southern part of the province, from a claim owned by two prospectors named Wollaston and Arundel. I met Wollaston and arranged to see the prospect. We went into the Similkameen by a rather circuitous route, down the Okanagan lake to Penticton, then to Fairview, where we were joined by Arundel, Wollaston's partner, and on to Twenty Mile Creek. I made a stay of about an hour and a half on the claim and sampled the showing. Intuitively it came to me that it had the earmarks of a mine. I sent the samples to a Montana assayer. His returns were encouraging. They looked too good, so later on I went back to the claim myself and resampled the ore. Again I received big results and that determined me to secure the property. I bonded the Nickel Plate, Bull-dog, Sunnyside and Copperfield claims for $60,000. Development started and the prospects in time became a mine. That delay at Victoria was responsible for my getting the Nickel Plate," he concluded.

Godenrath continued, "It was on the 12th of January, 1899, that a gang of 18 men started development work at the Nickel Plate. All supplies had to be packed in forty miles, and the early work of proving the ore body was prosecuted under the greatest disadvantages. As the work progressed Mr. Rodgers began to gather in other claims, and two months before the bond had expired on the Nickel Plate group, it was taken up and the balance of the money paid."

Early in the winter of 1898, M.K. Rodgers set about to prove the claims which he had bonded. The Similkameen generally has a pleasant climate, even in winter, but the Nickel Plate mine is located 4000 feet above the valley, buffeted by strong mountaintop winds and

Nickel Plate Mine site, 1901. Work on the Nickel Plate Mine, as all the claims bonded together became known, started early in January, 1899. The first work was done on the discovery showing of the Nickel Plate claim - elevation above sea-level, 5, 850 feet - and consisted of stripping off some of the overburden and driving a wide open-cut across the ore-body. This cut was blasted out to about an average depth of 9 feet. Pin-heads of free gold were frequently visible in the sides of the cut. The next major work was the driving of a cross-cut tunnel - elevation at portal, 5,750 feet - which was designated as No. 3 tunnel. This tunnel cut the ore-body about 400 feet in from the opening of the tunnel, but it was slow going owing to the extreme hardness of the rock. In 1901, a small Ingersoll straight-line air-compressor and steam-boiler were hauled in from Penticton over the newly built Nickel Plate road. The compressor was installed in the draw at the foot of the slope below the tunnel portal. This installation speeded up the driving of the tunnels. Nickel Plate Mountain was an ideal location for a mine. There was an ample supply of timber for immediate use, and nearby springs above the camp allowed the installation of a gravity water system. The southern slopes of Nickel Plate Mountain, and other mountains in the area, are more or less covered with a scattered growth of Douglas fir, among which bunch-grass and other grasses grow freely. Northern slopes are usually covered with mixed stands of fir, jackpines, spruce and balsam.

deluged with snow which does not melt until late spring. These were the conditions that Rodgers faced when he hired the local Indians to shovel a trail through the snow to the mountain and help bring in supplies. The supplies and equipment were for a crew of 20 men who began a tunnel on the claim. George Cahill and a partner, Yates, built a cabin for their accommodation near the Sunnyside claim.

The prospectors of "Hedley Camp" had to record their claims and assessment work at Fairview, where C. Lambly was gold commissioner for the Osoyoos mining district. News of the gold strikes spread quickly, especially in the Fairview camp where the claims had been recorded. Many miners came over to the new camp in the autumn of 1899. By 1900 almost all of the mountain had been staked and recorded. For some reason a fraction was overlooked and was still unstaked. George knew this, and once when he was strapped for cash he staked

This is the Nickel Plate Camp buildings to 1910. The early buildings were all of log construction, but in 1902 a new dining-room was built and was of frame construction, with lumber hauled in from Penticton. This was a two-storey building abutted on to the log-built kitchen, which had served as both dining-room and kitchen. The top storey was designed for a hall or sleeping-quarters. This building was known as "Cameron Hall". Directly across the road was a large two-storey bunkhouse, which was always known as the "Hot Air Tenement". A little farther down the track was the log house of the mine foreman. Beyond that was the two-storey store building with basement, partly of log and partly frame construction, built about 1901. Lumber for this building was also hauled in from Penticton. Almost directly across from the mine foreman's house on the upper side of the road stood a log building, the front part of which was used by the mine engineer as a draughting-room, and the rear as his sleeping quarters. Further down the lower side of the road was first the manager's two-storey log house, and then followed three smaller log houses for employees. These last four buildings stood on the Sunnyside claim. Back toward the cookhouse and going up the road towards No. 3 tunnel, there stood another small log cabin used for a time as a school-house. In the summer of 1905 four additional frame cottages were built for the employees. The lumber came up from Hedley over the company's tramways. These cottages were built in a row near the northern limit of the camp, and the little settlement went by the name of "Bogus Town." Nearby stood the company's stables and the first compressor-house. Doug Cox photo collection

the claim, not for himself, but for a Mr. Duncan Woods who had moved from his ranch near Trout Creek to Hedley. He reportedly paid $15.00 for eight and six tenths acres, that was called the Mascot Fraction. This "Mascot" was later to become one of the richest gold producing areas of British Columbia! "Dour old Dunc" held on to that property and would not sell to the original company at any price. He eventually did sell to a later company for a reported $80,000.

Supplies for men employed at the mine site and camp, plus construction materials, were first brought in from Fairview. George Cahill was in charge of the first pack train which left Fairview in November, 1898. Thirty-five horses loaded with supplies made their way from Fairview to Hedley Camp. Later, freight was hauled from Penticton to Keremeos, and then taken by packhorse up to the mine site at Nickel Plate via the Camp Rest Trail. In August of 1899, M.K. Rodgers made a journey to Vernon, where he met L.A. Clark who agreed to survey and build a road from Penticton to Nickel Plate.

Lenard Clark was born in the state of Vermont in 1840. After fighting in the Civil War he moved to Ohio where he married and started a family. He and his family crossed the plains in a covered wagon, arriving safely in Spokane, Washington. He had had experience with irrigation work, and did work as a railroad grade contractor in Washington. After his livery stable and store burned in Northport, he came north to Calgary, where he put in an irrigation system for W.C. Ricardo. Ricardo later moved to Vernon, where he became manager of the Coldstream Orchard. He sent for Clark to supervise installation of the irrigation system for the company.

In 1898, L.A. Clark went over the Chilkoot Pass to the Klondike, and presumably returned to the Vernon area in time to meet Mr. Rodgers, who engaged him as a contractor to build a freight route to the Nickel Plate.

The Public Works Report of 1901, referring to the Nickel Plate Road, is as follows: "Route surveyed and road built, (about 5 miles is as yet only 5 feet wide) by M.K. Rodgers, assisted by a Government grant of $4,000.00. This road starts at a point on the Penticton/Fairview Roads 2-3/4 miles Southwest of Penticton, and runs southwesterly through a farming, grazing and mining country, a distance of about 30 miles, to Nickel Plate Mine. 125 miles of trial lines were surveyed in order to provide the proper route:

> Forest cleared . . . . .79,200 feet, 33 feet wide
> Grubbed . . . . .79,200 feet, 10 feet wide
> Graded . . . . .68,835 feet, 10 feet wide
> Corduroyed . . . . 125 feet, 16 feet wide
> Cribbed . . . . .1,500 feet, 8 feet wide
> Excavated earth . . . . .129,360 feet, 3 feet wide
> Excavated rock . . . . .1,000 feet, 9 feet wide, 10 feet deep
> Filled in . . . . .12,936 feet, 4 feet wide, 4 feet deep
> Built 10 bridges, averaging 40 feet by 16 feet, by 8 feet.

Lenard Clark with his faithful horse, Jock, surveyed the original road to the Nickel Plate Mine.
Clark measured distance by counting measured wheel turns by a rag tied to the spoke of the wheel.

Clark Estate photo

3. Developing the Mining Complex

In 1902, when sufficient ore was found to justify a mill, it was found that the original charter granted the Yale Mining Company by the British Columbia government was inadequate to cover the construction of tramlines, power flumes, and a reduction plant. An application was made, and in 1903 a British Columbia charter was obtained, resulting in the formation of the Daly Reduction Company. The original company was retained and used as a holding and operating company. There was no water power on the mountain close to the mine to power the reduction plant that was necessary to extract the gold from the rock. A railway and tramline had to be built to haul the ore to a location where water power was available. The closest source of water power was on Twenty Mile Creek, in the valley about 4000 feet below the mine.

One of the first projects in 1902 was to make preliminary surveys for a water flume and a tramline. By early spring of 1903, construction of the flume to supply water power to the mill was begun, and also, the tramline and stamp mill were built at this time.

The single construction of any one of the three, the flume from Twenty Mile Creek, the tramline and skip, or the mill and concentrator, would have been extremely difficult, but to attempt all three at the same time was a monumental task! Each project was under a contractor or supervisor. Mr. Munson of Grand Forks was given the contract for the dam on Twenty Mile Creek, and the flume from the dam to the mill. James McNulty was the contractor for the mill. The skip, and the tramline were installed by a Swedish contractor from Grand

The ladies of Hedley often met in the great outdoors to socialize. However, the beverage was served in matching china cups and saucers. Please note "Hedley" crocheted on the tablecloth and the tent set up behind the women.   Herb Clark Estate photo

Tunnel and catwalk on the Twenty Mile Creek flume. The walk was used to bring in supplies and also protect the flume from falling rock. The flume would supply motive power for the mill at a point 3 miles distant from the mine (capacity of 1000 inches of water) to supply power to run the mill, generate electricity, and compress air. Mountain goats in the area were eliminated as the goats often loosened rocks which shattered the boards on the flume. The goats have since returned to the area.   Herb Clark Estate photo

Forks.

The water flume was built by J.K. Fraser, whose main leveling instrument was a two foot hand level and a twelve foot straight edged board! The heavy timber, which was milled locally, had to be hauled in by teams and then carried by hand to where it was to be used. The wider the lumber, the fewer the cracks to leak! A cat walk on the top of the flumes served two purposes; firstly, it was a way to bring in the lumber for further construction and steel for drilling the rock, and secondly, it prevented the flume from being damaged, or at least it slowed down the rocks that fell from the cliff face. There were many mountain sheep in the area, which, unfortunately, had to be exterminated, as the sheep would loosen rocks which crashed through the flume.

The waterway had to be level, reasonably straight and attached to the mountain in some way. Mother Nature had not taken this flume line into consideration, and consequently large boulders had to be removed, and tunnels drilled and blasted. In some places arrangements had to be made to suspend the creation from the cliff face. The boulders were moved by means of a "gin pole", a derrick-like structure to which a block and tackle was attached and powered by a lot of muscle and elbow grease. In order to drill rock by hand to make the tunnels, there had to be two people. One person held the steel, a two foot or more length of round steel with a chisel end designed to cut into the rock and break small pieces loose. The other member of the team swung a large hammer, which drove the steel into the rock. After each

Clearing the right of way and building trestles for the flume from Twenty Mile Creek. Water from Stray Horse Lake was the source of Twenty Mile Creek, and consequently the source of water power for the Reduction Plant. Ditches were dug to drain water into the lake and a steam engine was installed at the lake to pump water into the creek to maintain a steadier flow of water.

Herb Clark Estate photo

blow the steel man turned the steel a quarter turn to allow for a new steel cut in the rock. It was an extremely slow process. After the hole was made it was loaded with dynamite, referred to as "powder", and blown. When the smoke settled, the fragmented pieces of rock were removed and the process started again. This slow process was actually used to drill three tunnels, plus holes had to be drilled into the cliff whenever the cliffs were vertical to give support to the flume at that particular point. The flume went some three miles back to tap the headwaters of Twenty Mile Creek.

The mill was contracted by a Mr. McNulty. The concentrator was located on a hillside to utilize gravity in the milling and concentrating of the raw ore. The mill's foundations were of

Removing boulders from the flume line with a gin pole, (1903). A gin pole was a tripod and hand operated winch system, which was used to lift the boulders from the flume bed and move them to the side. Judging by the size of the boulders, the apparatus was highly successful. Please note the long skirts of the lady who has come to view the progress.   Herb Clark Estate photo

stone and cement, and the construction was of heavy timber beams.

Timber and lumber were supplied by Messrs. A.R. Tillman and J.J. McDonald who brought a portable sawmill and planer from Phoenix, and located across from the mouth of Sterling Creek. The first mill was set up in August, 1901, in 1902 they brought a new plant and set up on the other side of the river near J. Neills' Ranch, where there was more suitable timber. In 1903, Tillman sold his interest to Angus Stewart.

Harry Barnes, an employee and official of the mining company, was also a historian. He left us a very accurate and detailed account of Hedley Camp and of the early Nickel Plate Mine. He was also an excellent photographer. In his article, "Early History of Hedley Camp",

The Yale Mining Company and Daly Reduction Co. staff, July 15, 1907.   Pat Wright photo collection

he gives us an insight into obtaining supplies and equipment.

"By the late fall of 1903 the mill building was about completed, and a good start had been made on the installation of the machinery. The tramways were all graded and most of the track laid, and also good progress had been made in the building of the Twenty Mile Creek flume. It might be interesting to note that the ore-crushers and stamp-batteries were made in eastern Canada, as was also the large air compressor for the new power house. But the ore conveyors, Frue vanners, waterwheels, pumps for the cyanide plant, electric locomotives for the tramway, and most of the electrical equipment, came from the United States, as did also the twenty large tanks for the cyanide plant. Twelve of these tanks were 34 feet in diameter by six feet in depth, and the remaining eight were 30 feet in diameter by 10 feet in depth. All were made from California redwood, which had been knocked down and shipped from San Francisco by boat to Vancouver, then by Canadian Pacific Railway to Okanagan Landing. From there they were transferred to the lake steamer and delivered to Penticton, from where they were hauled by freight teams the 50 miles to Hedley."

On May 14, 1904 the Similkameen Star newspaper reported that,

"The stamp mill made a trial run last week - a battery of ten stamps working nearly an hour. Some of the shoes came off and a stop was necessary. The full forty stamps will soon be dropping continuously and then gold bricks will be rolling out."

On July 9, 1904 the same paper reported that all 40 stamps were in operation.

The tramline from the ore bin station to the tipple covers a distance of about 10,000 feet and 3,300 feet in elevation. The gravity type tramway was designed to operate in two sections

Pay Roll of *The Daly Reduction*

For the Month Ending ___ NOV __ 190__

No.		NAME.	HOW EMPLOYED.	No. Days.	Rate per Day.		Amount Due.		DEDUCTIONS.						
									Board.		Merchandise.		Hospital.		Advance.
					Dols.	Cts.	Dols.	Cts.	Dols.	Cts.	Dols.	Cts.	Dols.	Cts.	Dols.
	1	E M Mills	Carpenter	6	4	00	24	00			23	00	1	00	
	2	H. Mills	Laborer	5	3	00	15	00			14	00	1	00	
1008	3	Wm Comstock	"	5½	3	00	16	50	6	00			1	00	
1010	4	Wm Summers	"	9	3	00	27	00	9	00		40	1	00	
1133	5	Percy Bragg	"	6½	3	00	19	50					1	00	
760	6	E. A. Hill	"	7	3	00	2?	90	8	00				50	
	7	A. D. Robinson	"	7½	3	00	21	75	8	00	6	00	2	00	
1131	8	Wm Mason	Mason	5½	6	00	33	00	6	00			1	00	
1136	9	J. R. Way	Pipe Man	7½	4	00	30	00	2	00		55	1	00	
1137	10	Geo Slack	"	11	3	00	33	00	12	00			1	00	
	11	J. Senson	Trestle Work	14	3	00	70	00	14	00			1	00	
	12	J. Thos McQalkin	"	13 oths	3	50	46	02	14	00		45	1	00	
1191	13	J. E. McCallery	Laborer	6½	3	00	19	50	7	00			1	00	
	14	Wm Kirby		20½	3	00	63	35	22	00	27	35	1	00	12
	15	P Brodhagen	Foreman	24⅝	4	00	98	00	25	00	7	75	1	00	
	16	Jas Murphey	Laborer	20	3	00	60	00	25	00	10	70	1	00	
	17	O. Borgenson		24¼	3	00	72	75	25	00	3	30	1	00	
	18	I. L. Oran		24¼	3	00	74	25	25	00	7	40	1	00	
	19	H. Swan		17½	4	50	78	75	18	00			1	00	
	20	E. Morten		21¾	3	00	65	25	21	00			1	00	
	21	H. Baines		25	3	00	75	00	25	00	4	15	1	00	
	22	H. D. Sauve		26¼	3	50	91	87	25	00	9	50	1	00	
	23	M. Hanson		21½	3	50	90	25	25	00	7	25	1	00	
	24	R. Dickson		24¾	3	00	74	25	25	00	10	65	1	00	
	25	E. S. Weaden		20	3	00	60	00	20	00	13	75	1	00	
	26	H. W. Yates		25¾	3	00	77	25	25	00	23	50	1	00	
	27	Ah Cheon	Cook	9	1	00	9	00					1	00	
	28	Ah King	"		1	00	21	00							
	29	Tom Sing	"	1 month			55	00			3	15	1	00	
	30	Chas Yung	"	1 month			50	00			1	75	1	00	
	31	Ah Sing		3½	2	25	7	87			5	75	1	00	
1356		P. F. Bragg		⅔			60	00			22	15	1	00	
							1559	91	401	00	201	40	30	50	12

Stone 9/c 60.00
Bdhouse 9/c ... 142.87 } E963
.... 63.00
Pipeline 1294.04
Gen Labor
$ 1559.91

Certified Correct: ___________ Foreman.

Employees at *Inc Tram*

		Total Deductions.		Balance Due.		We, the undersigned, acknowledge to have received the amount set opposite to our names.	REMARKS
Cts.	Dols.	Cts.	Dols.	Cts.	Dols.	Cts.	
		84 00				Voucher Signed	
		15 00					
		7 00	9 50			Voucher allowed	1003
						B & B 75	1300½
		10 40	16 60			CB61 allowed	
		1 00	18 50			CB65 do	
	1	8 50	13 40			CB47 No allowed	
		15 00	6 75			CB H L Robertson	1285½
		7 00	26 00			CB65 Voucher allowed	
		13 55	16 45			CB65	
		13 00	20 00			CB65 do	
		15 00	55 00			CB75	1286
		15 45	30 57			CB75 T. J. McAlpin	1287
		8 00	11 50			McPaulig	1191
00		6 25				Wm Kirby	
		33 75	64 25			Cr o/c EJ66	
		36 70	23 30			Cr a/c EJ66	
		39 30	43 45			CB75 J Burgess	1288
		33 40	40 85			CB75 S Doran	1289
		18 00	60 75			CB75 A R Sway	1290
		22 00	43 35			CB75 E S Mitton	1291
		30 15	44 85			CB75 H D Barnes	1292
		35 50	56 37			CB75 H D Sawry	1293
		33 25	57 00			CB75 Hans Hansen	1294
		38 65	37 60			CB75 R Dickson	1295
		33 75	26 25			CB75 E Wieden his Mark witness L Quick	1296
		49 50	27 75			CB Cr a/c EJ66	
		1 00	8 00			CB75 Lockrey witness F G White	1297
						King witness F G White	1298
		4 15	50 85			Wm his mark witness F G White	1299
		2 75	47 25			E his mark of Lloyd witness F G White	1300
		6 75	1 13			King witness F G White	1301
		23 15	36 85			J F Bragg 1206	1206
00		644 90	915 01				

Approved : _________________________________ Superintendent.

The Hedley Gold Mining Co. Ltd. & Daly Reduction Co. staff, 1912. Back row, L to R: H.G. Freeman (storekeeper); Wm. Sampson (Mine foreman), Dickenson, I.O. Merrill (son of president), H.D. Barnes (Purchasing agent). Front row L to R: Arthur Clare (mill foreman), Gilbert Eiken (electrician), Rosco Wheeler (mill superintendent), B.W. Knowles (mine superintendent), G.P. Jones (General superintendent), S.L. Smith (with beard) (accountant), E.H. Williams (assayer), Isaac Merrill (president).   Pat Wright photo collection

with a slight deflection at the mid-station.

According to Carl Loomer who worked for the mining company, as well as his father before him, the first cable was taken to the top to the Nickel Plate, and then was reeled off the reel by pulling the end of a cable down the line with a horse. Subsequent tram cables for the lines were brought to the bottom of the skip, where they were towed up behind the ore car to central, where the cable was put around the drum and put in place. A similar procedure was used to change the cable on the top section from central to the ore bin. The cable was changed once each year so that passengers could ride on the skip, regulations concerning these cable cars being very stringent.

One important feature of the mill site was the "compressor house" or "power house". This building was located to the west of the mill site and contained the pelton wheel, a sophisticated water wheel that powered a dynamo and supplied electricity. The building also housed two steam boilers that were used as auxiliary power when the flume from Twenty Mile ceased to flow.

A compressor was located here also. The compressor located at the bottom of the mountain supplied air to the mine through six inch pipes which followed the tramline. It was more economical to compress the air at a lower elevation, than to compress the rarified air at the mine site. The tramline was about 10,000 feet long, and the mines were located about a mile

Freight teams hauling timber for the construction of the Daly Concentrator. Note the tie down system on the freight wagon. Other than lumber, which was milled locally, all supplies and machinery had to be freighted in from Penticton by four-horse and six-horse teams. The distance from Hedley to Penticton is 50 miles, and half the distance was over what was then a rough mountain road with many steep grades. Penticton to the Nickel Plate mine is about 35 miles, but this was all a rough mountain road with even steeper grades and more switchbacks than on the Hedley road. The nearest telegraph office was then at Vernon, two days' travel from Hedley.

Herb Clark Estate photo

and a half from the ore bin; the pipe ran all the way, supplying fresh air to the miners, and compressed air to operate pneumatic rock drills.

Before this air line was installed, the compressed air supply for the mine was produced by a steam compressor located at the mine. The steam plant and compressor were located in the draw below the bunkhouses at the mine site. The barn for the company horses was located farther down the valley.

In the early picture of the mine taken in 1908, there is a tailings pond in the foreground. These tailings were pumped back in when a later company, the Kelowna Exploration Company, took over. The building to the right contained coal, gravel and coke; next was the machine shop. Later a welding shop and electrical shop were built in the foreground. The wood in the foreground is not for ties for the tramline. Used timber from the mine which had been used as shoring was used for ties, this timber was brought down from the top.

Cordwood was brought in by the carload, and was used to heat the lime kiln. There was a limestone bluff just south of the mill which produced the raw limestone. A man by the name of Fred Lyons worked the kiln. It took about 20 cords to complete a "burn" at the kiln. Since the mill did not run during the winter this was Mr. Lyon's job, to stock the wood, which he

Construction, early 1904. A British Columbia charter was obtained for the Yale Mining Company, which became the holding and operating company. When it was decided to build a mill at Hedley, it was found that the Yale Mining Company's charter was not broad enough to provide for the building of tramways and power flumes, nor for the expropriation of land for rights-of-way. A second company, the Daly Reduction Company Limited, was incorporated, and a British Columbia charter obtained for it early in 1903. It became the operating company for both the mine and the mill, the Yale Mining Company existing only as a holding company. Marcus Daly died while the mill was under construction and ownership of the Nickel Plate then became vested in his estate.

Herb Clark estate photo

usually did in cord stacks, which gave him some idea of the amount of wood he had. This project was first started by his brother, George, and then transferred to Fred.

The slaked lime was used in the mill to make the ore settle faster, by creating an alkaline mixture, thus speeding up the process. The quick lime was found not to be as high a grade as could be purchased elsewhere, consequently the company eased the lime kiln out of operation. This was done during the time when the mill was under the management of Kelowna Exploration Co.

Carl Loomer went on to describe the water flume: "The man that built that flume, built it on an ordinary two foot level. There was no survey instrument or nothing used on that. And he ran that a quarter of an inch in 12 ft. and it must've worked, because it never flowed over!

The old original road up to the Nickel Plate Mine from Green Mountain, started at LeLievre's place, and I've forgotten how many switchbacks were on that, and sand! When they eventually got cars that would go up there, boy, it was a treat to sit in the back seat and

One of the first photos of the Daly Reduction Plant at Hedley. The building in the lower left of the photo housed a steam powered air compressor.   Herb Clark Estate photo

bump your head, and the throttle daren't ease up or you'd get stuck in the sand. I took a Model T Ford up there. The mine was down in 1925, they had a fire up there, and I took this old Model T Ford. They had in those days what was called a Ruxsell axle, in the truck it was called a Fuller axle. The thing went out halfway up and there we sat all night. The manager of the place, a man called G.P. Jones, came up to see what had happened and he lowered me down over the road with an Ajax Nash, a rope tied between us. If anything had happened to his brakes we'd have both gone over the dump; there wouldn't have been anything holding either one of us, because I had no way of stopping the auto I was driving.

The fellow who built the Twenty Mile Creek flume was J.K Fraser. There are three tunnels between the mill and the dam on Twenty Mile Creek. Those tunnels were all made in the day of hand drills. That's all the power there was, nothing else, one man swinging a hammer, the other holding the steel. Sometimes you'd get a quarter turn, sometimes you'd get a full half turn. If the man on the hammer was a good man, you'd be damned fast to get more than a quarter, because you were also watching your nose with this big hammer coming by you.

There was no compressor at the mine. All the air was compressed down here. It went in a 6" pipe from here up to the mine. This end of that first cable tram is 5500 feet; the next one is 3800 feet. Then you had a mile and a half to go to the mine. That was a 6" line from two compressors here at the bottom. Due to elevation and, of course, the expense of getting it there, the difference in elevation of air meant they could produce it cheaper here than they could up there. The elevation had a tendency to work on it."

The Daly Reduction Plant in June 1908. The early operation of the plant was unable to extract all of the gold from the processed ore. This test slag pond was a temporary measure, and the ore was later reprocessed as the plant's technology improved. The massive cordwood pile was used to slake lime, which helped precipitate the gold ore in suspension in the tanks. The timber was for the mine.   Herb Clark estate photo

The first reduction mill operated by the Daly Reduction Co. was powered by Pelton wheels with water delivered to the mill by the Twenty Mile flume. This plant reflected the almost textbook ore reduction techniques used by the industry at that time. In 1909 the new owners, Hedley Gold Mining Co. Ltd., utilized the same water power to generate electricity. Electric motors then powered the equipment in the mill. The steam boilers, which were used as auxiliary power, were increased.

Larger ore crushers and tube mills for grinding the ore were introduced in 1914 and 1916 to increase the efficiency of the mill, and compensate for the hardness of the ore that was being extracted at the Nickel Plate Mine. In 1932, when the mine and the mill were taken over by Kelowna Exploration Co. Ltd., this machinery was upgraded with more efficient ore crushing and ore recovery equipment.

The next few pages are rather technical, and describe the workings of the early ore reduction mill at Hedley, and will possibly appeal more to the avid miner than to the general reader. To the miner, please enjoy, to the general reader, please enjoy the photos and captions. The formal names of some of the now-antiquated milling equipment refer to the man or the company that manufactured the machinery.

Harry D. Barnes described the workings of the Daly Reduction Company from 1904

The Daly Stamp Mill and Cyanide plant at Hedley in 1912. The three men are: left, Mr. Dickenson, director; center, Mr. I.L. Merrill, president; and G.P. Jones, superintendent, on the right.

Pat Wright photo collection

to 1909. "The primary crushing plant of the new mill consisted of two Farrel Jaw crushers, 20x10 and 20x6, and ore conveyors. The ore was pulverized by the forty stamps of the stamp mill. Each sorter of the stamp batteries discharged its product onto a series of copper plates for catching the free gold. At stated intervals the stamps of a battery would be hung up and the plates dressed, which meant that the amalgam would be removed and fresh quicksilver added to the plates. Power for the stamp mill was furnished by a Pelton impact water wheel coupled onto the end of the line shaft. A second water wheel installed nearby furnished power for the crushers and conveyors, the power being transmitted up to the crusher floor by means of a rope drive.

The crushed ore from the stamp batteries, after passing over the copper plates, was conveyed by overhead launders to the vanner floor, and there distributed to each of the 24 vanners. The vanner concentrates were pulled each day and dumped through trap doors to the bin below. After a period allowed for drying, the concentrates were put into double sacks, a heavy cotton duck sack on the inside and a strong jute sack on the outside. The sacked concentrates weighing a hundred pounds or more per sack were hauled by wagon to Penticton, and from there shipped to the Tacoma Smelter. The tailings from the vanners were carried by launders down to the cyanide plant below for treatment there. A third water wheel installed on the battery floor level, furnished the power for driving the vanners."

Cyanide is a chemical which will dissolve gold and silver from the mineralized ore, consequently the substance is used extensively in the gold and silver extraction process.

"The cyanide plant was equipped with 12 sand vats of 34 ft. diameter x 6 ft. staves, 4 conical bottom slime vats of 30 ft. diameter x 10 ft. staves, 2 gold tanks of 30 ft. diam-

Riffle boxes were used to separate the coarse gold. In the early days of the Nickel Plate Mine there was visible gold in the high grade ore areas. Slurry, ground ore and water, poured over the slats in these boxes, and the gold was trapped behind the slats, similar to a sluice box in placer mining.
R.N. Atkinson Penticton Museum photo

eter x 10 ft. staves, and 2 sump tanks of 30 ft. diameter x 10 ft. staves. These tanks were all made from California redwood, and supplied by the Pacific Tank and Pipe Co. of San Francisco. The two centrifugal pumps were supplied by Knight Bros. of Sutter Creek, California, and each pump had its own individual water wheel. The Knowles vacuum pump was belt driven by a third small water wheel.

The gold bearing solutions from the strong gold tank were piped and flowed by gravity into the zinc boxes. The barren solution flowing out of the zinc boxes was discharged by gravity into one or the other of two sump tanks. The zinc shavings for the zinc boxes were cut at the plant on a small lathe, and the sheet zinc was usually bought in carload lots. The zinc boxes were cleaned at regular intervals, and new zinc shavings put in. The gold precipitates from the zinc boxes were put into a small tank about 6 ft. diameter - 5 ft. in height, and there treated with sulfuric acid. After treatment in this tank, the precipitates were ready for roasting in the refinery below. The equipment in the refinery consisted of a retort, a small reverberatory furnace for roasting the precipitates, and two coke fired furnaces for melting the precipitates and retorted gold. It was the usual practice to make a cleanup once a month, and two gold bricks, one from the free gold caught on the plates, and the other from gold recovered in the cyanide plant, were taken out under special escort to Penticton, and from there shipped by Dominion Express to the U.S. Assay Office at Seattle. About 1908, the amount of free gold recovered on the plates each year began to get progressively less, and in about 1911 or 1912, amalgamation was discontinued and the copper plates taken out.

The equipment in the power house consisted of a Westinghouse 90 K.W. A.C. gener-

Filtering the slurry, which is ground ore and water. After the leaching process, when cyanide was added to dissolve the gold, the solution was the so called "pregnant solution" because it contained the gold and silver. This is basically the same process as is used today. The "Merrill Croll" method adds zinc oxide to the pregnant solution,which makes the gold and silver precipitate out as a solid again and can be caught on the presses. This is the product put into the furnaces to produce gold buttons or ingots.

R.N. Atkinson, Penticton Museum photo

ator, driven by a Cassel water wheel operating from a 420 foot head. This unit furnished the current for lighting and for operating the electric tramway at the mine. A large Canadian Rand cross-compound air compressor, driven by a Pelton impact wheel, also operating under a 420 foot head, furnished compressed air for the plant and mine. In the adjoining boiler house, a Mumford boiler and a Vancouver Engineering Works horizontal, tubular boiler furnished steam for heating the plant in winter time. Cordwood was used as fuel for heating the mill.

In the Spring of 1909, the Daly Estate gave an option on all of their holdings in the Hedley Camp to a New York syndicate headed by the late I.L. Merrill, and a party was sent in by them to make an examination of the Nickel Plate mine. The results of the examination proved satisfactory, and their option was taken up in September of 1909. The new owners organized and obtained a British Columbia charter for the Hedley Gold Mining Company Limited, and this Company now became the holding and operating company. Gomar P. Jones, who had been mine superintendent at the Nickel Plate, was made General Superintendent, and Roscoe Wheeler of Oakland, California, was engaged as Mill Superintendent.

About the first change made to the plant by the new company was the installation of an auxiliary steam plant, and the replacing of water wheels in the mill with electric motors. To furnish the required power, a Canadian Westinghouse 350 K.V.A. A.C. generator was installed in the power house, which was driven by a Doebel water wheel

The auxiliary steam plant was situated near the mill and consisted of three return tube steam boilers and a Goldie-Carliss steam driven engine directly connected to a 375 kw generator. Blakeburn coal was burned in the unit, which often supplied electrical power when the river was low or frozen.

Estabrooks photo

This machine shop, located at Hedley and completed prior to 1905 by mine manager M.K. Rodgers was crucial to the mine and mill operation. Any needed repairs to the plant were manufactured in this up-to-date facility. A central power shaft supplied the power and belts worked off this shaft to the individual machines. Operators often moved the belt off the pulley with a long stick when a machine was not in use.

Dave Innis photo collection

In 1914 a 24 x 36 Traylor crusher was installed. In the background is the ore tram from the ore dump to the mill. The cost of the Traylor crusher was $7,079.54, the cost of a house built at the same time for the mine engineer was $1979.34.   Estabrooks photo

operating under a head of 420 feet. A Goldie and McCulloch condensing engine was attached to this unit, and was used as an auxiliary in winter time. A new penstock was also laid from the forebay of the Twenty Mile Creek flume in the power house. The air compressor was rebuilt and an auxiliary cross-compound condensing steam unit added to it. A battery of three horizontal tubular steam boilers, made by the Jenckes Machine Co. of St. Catharines, Ontario, was installed in the boiler house. Princeton coal, shipped down over the Great Northern Railway, was used as fuel.

Harry Barnes continued: "As greater depth was attained at the mine, it was found that extraction of the gold content of the ore became increasingly more difficult. To obtain a better extraction, it was decided to resort to fine grinding and direct cyanidation. In February of 1916, orders were placed for four 5 ft. x 22 ft. Allis Chalmers wet grinding tube mills, five Dorr duplex rake classifiers, and other miscellaneous equipment. It would have been probably late in 1916 before this equipment would have been installed, and it is likely that the changeover to fine grinding and direct cyanidation was not made until 1917. Under the new system of milling, the stamp batteries worked in a closed circuit with the tube mills, and crushing in the batteries and fine grinding in the tube mills was done in solution instead of water as heretofore. The pulp off the Oliver was pumped back to a tank above the vanner floor and then concentrated on the vanners and Deisters. Concentration now became the final process of extraction."

Joe Bromley recalled that the tube mills were 5 1/2 by 22 feet long. "When they first started using those tube mills, they used to get flintstones from Norway. There were five of the tube

The addition of a new 24 x 36 inch Traylor jaw crusher added to the mill equipment. Judging from the cables which are holding the unit and the triumphant pose of the men, removing the massive steel unit and installing it in the mill must have been a momentous feat.　　　Estabrooks photo

mills. You had to feed every shift, there were so many flintstones in there, and they ground the cuttings from the batteries, or stamps, then liquid carried these cuttings down into these tube mills. It was ground up finer, by the pebbles in there. They found out later that the ore from the Nickel Plate was alright to use for grinders. Then they had a man for day shift picking them off the belt. They had two crushers, a small one and a big one. The rocks were six inches and a man stood there all day picking them off the belt. They collected about five wheelbarrows a shift. They fed five wheelbarrow loads to each tube mill every eight hour shift."

Possibly the workings of the Daly Reduction Mill until it was taken over by Kelowna Exploration Co. Ltd., can be best described by going directly to the "Report of the Ministry of Mines, 1929" to a description of the mill supplied by Rosco Wheeler, the mill superintendent at that time.

"The ore is conveyed to the tipple above the mill. From the tipple it is trammed about 300 feet to the crusher-bin, from which all the ore passes into a Traylor-built, swing-jaw crusher of 36 by 24-inch opening. The entire product from this first crusher drops to a belt, which delivers its load onto a set of grizzleys 3 feet long. The oversize from these grizzleys goes to a second swing-jaw crusher of 20 by 10-inch opening. The product from this second crusher goes over a set of grizzleys 8 feet long, the oversize from which goes to a third swing-jaw crusher 20 by 6 inch opening. The third crusher drops its product to the same belt that carries all the undersize from the grizzleys and conveys it to the ore-bin that supplies the stamps. The maximum size of lumps delivered in this bin by the crusher is about 1 by 1 by 2 inches. All three crushers are placed in tandem. This battery-bin contains, when full, about four days' run for the forty 1,050-lb. stamps.

1914. Freighting in one of the four 5 1/2 by 22 foot tube mill. This tube is being brought from the railroad siding by a 4-horse team. Tube mills were added to the mill after the initial construction to improve gold extraction.   Herb Clark Estate Photo

From the stamps the ore passes first through a 12-mesh screen and then goes to five Dorr drag-line classifiers. The fines passing off the lower end of these are conveyed directly to the slime-tanks, while the coarser particles leaving the upper end go to five tube-mills, each five feet in diameter and 22 feet long, which work in a closed circuit with the classifiers. Eighty percent of all material leaving the lower end of the classifiers will pass through a 200-mesh screen. All crushing is done in cyanide solution. The lump lime is fed in the battery-bin and the cyanide solution first comes in contact with the ore under the stamps.

From the lower end of the Dorr classifiers the slimes go to three settling-vats of 30 feet diameter with 12-foot staves and conical bottoms of 22 degrees. This incoming slime settles readily, and the solution is drawn off continuously and goes to a lower sump-tank. When the charge of slimes in a settler amounts to the equivalent of about 100 dry tons, the slime feed is switched to the other two settlers. The solution of the slime charge in the cut-out settler is decanted as far as possible, then the tank is filled with barren sump solution, and the charge agitated and thrown over by a 6-inch centrifugal pump to an empty vat. Here the charge is agitated for about eight hours, or as long as time will permit, allowed to settle, and the solution decanted. The vat is then filled again with barren solution, reagitated, and the charge transferred to a slime-vat that acts as feeder for the Oliver filter-press.

All solutions decanted from slime charges go to a common lower sump-tank, and are strengthened by addition of lump sodium cyanide. From here the solution is pumped to a tank above the stamp-battery floor, and becomes the feed solution for the ore under the stamps. The overflow from this battery-feed solution-tank goes to 2 filter-tanks, 34 feet diameter by 6 feet high, with a canvas bottom. This clarified gold-bearing solution from the 2 filter-tanks is lowered to another tank, which is a unit of the Merrill zinc-dust precipitation system. Six hundred tons of this solution are sent through two 21-frame and one 11-frame Merrill precipitation-presses. Clean-ups from these precipitation-presses are made about three times per month. The zinc-dust sludge is treated with sulfuric acid, washed, dried, roasted, melted at the mill, and shipped as gold bullion.

The slime from the Oliver feed-tank goes to two 8 ft. long by 11 1/2 foot diameter Oliver filter-presses. The tailings from the filter-presses are elevated to a series of eight Spitzkastens. The overflow from these goes to ten Deister concentrating-tables and the middlings from these ten tables go to two other Deister slimers below the first ten. The overflow from the eight Spitzkastens goes to twenty-four 6-foot Frue vanners. From the Deisters and vanners the tailings are passed out of the mill. The concentrates are shipped to Tacoma, Washington.

The tonnage of ore treated per month is about 6,000 tons. Sodium cyanide used is 2 lbs per ton of ore. Lime used for all purposes is 8 lbs per ton of ore. Extraction varies with fineness of crushing from 88 per cent to 92 per cent."

Gold brick, poured June 1918. The arsenic pyrite in the ore left the gold bricks black. Consequently, the ore was sent to the smelter at Tacoma, WA for refining. Various attempts were made to refine the ore at the reduction plant, but they were generally unsuccessful.   Pat Wright photo collection

Carl Loomer spent most of his working career with the Nickel Plate Mine. He did take some time out to be a bus driver, but went back to work with the mine. He was with Kelowna Exploration Co. Ltd. when the mine and buildings were shut down.

"The cable was brought up by three or four horse teams. It came right from the top down. They took the first cable up to the top of the Nickel Plate, but they never did it again. The first cable came from the top, from then on it went from the bottom, because you had ore skips operating then. You'd tie one end of the cable to an ore car, and you had a three inch plank or four inch plank, whatever it was to hold this reel from going too fast, and they'd make the skip pull it. Now when the skip got up to Central they put the cable around their double drum and tied it on the skip coming down and brought it in place. This is an endless cable. It's one cable with two ends, but the way it goes around the drum, they call it an endless cable. In other words, the loaded skip coming down took the empty up.

There is a popular story of 40 mules being used to haul cable to the top of the mountain. Not so! The last people that owned and operated this mine were Kelowna Explorations, the Vice-President's name was J.W. Mercer. That is where the story originated. He told a story which took place in South America where 400 mules took cable into a mine in South America. This story became exaggerated and applied to here.

Once they got that cable to what is called the ore bin on top of the mountain, the same

The interesting thing about this 1916 photo is not the man near the work tram, which was higher in the front in order to carry construction materials, or the left hand cable which seems to have a cable tram car removed for unloading. The unique figure is the lady, dressed in a fashionable long skirt and hat, waiting to catch the ore tram up to Nickel Plate townsite, just as a woman might have waited for a street car in Vancouver or any other cosmopolitan city.   Estabrooks photo

This was a work cable car raised in one end to facilitate loading construction materials. The man in the centre is Wallace Knowles, the mining engineer who came from Colorado. Wallace married one of the girls from the Daly family in Keremeos.   Pat Wright Photo collection

thing was done, but they came down with a horse pulling it down on this reel. They would have a brake on the reel. These cables came in very substantial solid reels, as they'd have to. One weighed a little better than five ton and one almost four ton. Everything was wood in those days. Now they ship this stuff on steel wheels and send the reels back. In the old days, the old reels were burnt, unless somebody wanted them. But that's the way the cable went, it came in to the top and down the hill; there was a horse pulled the first one down and they controlled it, and did not let it go too fast with these plank brakes as they called them, but from then on it went from the bottom. Now, they had a freight car on behind these ore cars. The men would take maybe three or four wraps of this cable; it was quite a flexible cable, put it in the freight car, and with a clevis hook, as they called it, put on the back they took that up behind an ore skip. And they would go very slow. The cable was changed once every year. Passengers were not allowed to ride on it if it was over a year old. They were very strict. This was, at that time, the only way to get up and down to the mine. There was no road from this side. And that road you see winding up that hill wasn't completed until 1937."

Sandy Brent, who was a trucker when the supplies were first brought into the rejuvenated mine, later obtained a job at the mine. He was single at the time, and had to get up and down to the mine each day. The "new" Nickel Plate road had not been constructed, so the men who lived in Hedley walked the Kingston trail each day before starting a shift as a hard rock miner. After shift the men could look forward to walking down again.

Not everyone walked all the time. The skip ran all the time, however, in order to establish credibility with the mine inspectors and reduce the possibility of accidents on the tramline. Employees were required to obtain a pass before riding the skip. The number of passes

A passenger about to take the second half of the journey down to Hedley. The cable around the ore car was a safety measure, and the tram line cable was replaced frequently as a safety precaution, and the fact that passengers also rode the tram cars.   Herb Clark Estate photo

issued was limited. The men got around this by remaining concealed near the upper ore bin until the skip had started down, then they all jumped on the ore car for the trip down!The hitchhikers got off the skip before it reached the bottom ore bin and kept out of sight of the supervisory staff at the mill, skirting across the bluff and then down to the town of Hedley.

Sandy reminisced about times when he and other men lived at the bunkhouses at the top. The "action" was usually down in Hedley during the evening, where there were at that time various hotels or parlours catering to the workers. A favorite, or one of the methods of commuting at that time, was to nail two cleats on a short board which was set on the center rail of the tramline. Sitting on the board and using an old pair of gum boots on one's feet, that could be jammed against the rail and act as brakes, the men could make a rather unique and swift trip down to Hedley! When the boys wanted to go home they caught the skip back up the hill. If one of the men didn't have time to manufacture a "seat" for himself, he would "borrow" a broom from the bunkhouse and ride the rails sitting on the broom.

Ed Brent, who was for a time assistant foreman on the tramline, recounted a grizzly incident when he and a crew were taking a cable that was being salvaged down the tramline. The crew had the cable down to Central the night before, and it was to be taken the rest of the way first thing in the morning. "We started out of Central, started out slow at first, then we got going a little faster and a little faster. Right away I knew we were going faster than we should. We couldn't jump going across the trestle. We were on a flat skip about 14 feet long and four feet wide. I was sitting beside Louis Oldenburg. Two fellows were crouched down riding on the front of the skip. I told them something was wrong, that we were going too fast and

Ore cars or "skips" at Central station. Skips were changed to a different cable system at this point to compensate for the 20 degree angle in the direction of the tram line. The extra flange on the front wheel was used in dumping the ore car.   Herb Clark Estate photo

started yelling 'Jump! Jump! Get out!' They rolled off the front, I can still see him, Vic who was wearing a black sweater, roll down the hill. I climbed over the side of the skip, which was difficult because of the speed, and threw myself out, head up the hill. I can remember seeing Louis, who was wearing a felt hat, which was bent up against his head, froze there. When I threw myself out, I missed all the nails in the scrub blocks, but there was still 6000 feet of 1 and 1/8th cable coming down like a bullwhip. I pushed one fellow who was stunned up a bank and then ran for the bush. I knew Louis was gone. The skip jumped the track near a big cut. Louis was smashed into the side of the cut."

Joe Bromley described Yorkie, who ran the skip at Central: "If he thought anybody was scared to ride down the tramway he would just turn it loose down there. One time I went down first thing in the morning, the first skip that went from Central down, and they had snow, about six inches during the night, and he just let that thing fly down there with a big cloud of snow flying up around me!

Yorkie had a few claims. He was a prospector. After he left that job, he worked around the mill at Hedley, then he was prospecting. He had some claims a couple of miles southeast of the Nickel Plate. He had some friends at Princeton backing him, but it never turned out. He was an old time rock driller. He was married with no children. They used to have rock drilling contests at Hedley on Labor Day sports days. The drilling was done by hand drills. It consisted of a team of two men. Yorkie was on a team, but I don't think he was ever a champion.

The skips ran over a trestle a short distance below the ore bin, and on the lower side of the trestle the grade flattened off some. This caused the cables to go away up in the air. There

Central station located about mid point on the tram line. From the ore-bin station at the head of the gravity tramway - elevation 5,400 feet. On account of its course being deflected 20 degrees at central station, the gravity tramway was built for operating in two sections. The distance from the tipple to the central station is about 5,600 feet, and from the central station to the ore-bin about 4,000 feet.

Herb Clark Estate photo

was the danger that the cables would catch on the sides of the trestle and throw the skips off the track, so an overhead frame was built with rollers to hold the cable down. They were only 30 inches above the top of the skips, so people coming down had to lie down flat to get under them."

Sandy Brent used the tram line and the skip on a regular basis.

"Harriet (Sandy's wife) and I went up on the skip lots of times. The trestle is about 25 to 30 feet high. There was a fellow, Knowles, who was running the mine at that time, and he was coming up on the skip. If you're coming up and you're going to stop, the cable flops; they had wings on each side to stop the cable from catching on the trestle. Knowles was in the ore skip, going on the trestle, and they started stopping for Central and the cable started flopping. It flopped and hooked over one of those wings to stop the cable. It took the skip and over she went right there! He fell about 25 to 30 feet I guess. He never got a scratch, and that big skip fell on the ground just above him. He went on up to the mine. Knowles went to Seattle one time and he put his suitcase down to unlock the door in the hotel, stepped backwards, fell over the suitcase and broke his neck. Killed him right there!"

In 1935 Kelowna Exploration Co. Ltd. replaced the original track on the surface incline with 30 lb. steel, and new haulage cables were installed to handle the 5-ton capacity skips. The friction brakes on the tram system were replaced by motor-driven drums which proved to be much safer. The tram line operated with these additions for twenty years until the mine and mill closed down in 1955.

Bernice Glenn, Alice Bromley and Miss Gordon, who lived in Hedley, at the Mascot Mine. This society event was written up in the Hedley column of the Penticton Herald, Thursday, July 15, 1937. "Miss Bernice Glenn of Bellingham is the Guest of her uncle and aunt Mr. & Mrs. J. Bromley".
Bromley photo

Maurice Budd who worked at the mine in the late 40's and early 1950 took this photo of an ore car approaching Central Station on the tram line. Here the skip would be changed to a different cable, and the car would continue the trip to the ore bin above the Daly Concentrator. Roffel/Budd photo

In deep snow the tram line was too steep to attempt any snow removal. The tram car simply moved through spraying snow as the ore car came up and the loaded car went down.Herb Clark Estate photo

The upper half of the tram line railbed leveled off and crossed a ravine on a trestle and then made the steep rise to the ore bin, which was the top terminal of the line. A wooden arch with rollers held the cable down, preventing it from flying up and landing back other than in between the rails. The diagonal planks were to keep the cable from interferring with the trestle.

R.N. Atkinson Penticton Museum photo

5. FREIGHTING WITH HORSES

Supplies for men employed at the mine site and camp, plus construction materials, were first brought in from Fairview. George Cahill was in charge of the first pack train which left Fairview in November 1898. Thirty-five horses loaded with supplies made their way from Fairview to Hedley Camp. Later, freight was hauled from Penticton to Keremeos, and then taken by packhorse up to the mine site at Nickel Plate via the Camp Rest Trail. Camp Rest trail started up the Nickel Plate Mountain at a point where the new Nickel Plate Road now begins. Below the bluff a trail switchbacks up a canyon to the top. The Camp Rest Trail goes through what was George Cahill's property. It was the original trail by which supplies were taken up to the mountain.

In 1899, M.K. Rodgers went to Vernon and met L.A. Clark, who agreed to survey and build a road from Penticton to Nickel Plate. At that time, road survey work was possibly more exacting than it is today. A road grade that was not too steep, and which could be constructed with a minimum amount of work was essential. L.A. did much of his work using a one horse cart which had a mark on the wheel to calibrate distance.

The grade and the route on the Green Mountain and Nickel Plate roads were as a

Businessmen travelling to Keremeos or Nickel Plate often stopped at the Green Mountain Stopping House. The livery barn is located on the left. L.A. Clarke's cart which he used to measure distance when surveying roads is in front of the barn. The building to the right, still standing today, was used to stable the stallions. The structure partly dug into the earth bank is the root cellar. The gentleman in the center with the suit and riding jacket has been identified as Francis X. Richter of Cawston.   Herb Clark Estate photo

Gerald Clark, the son of L.A. Clark who surveyed the first road to the Nickel Plate mine, was for a time a freighter. His 4-up team features heavy harness and the warning freight bells are on the off leader horse. He carries a spare neck yoke on the front of his narrow rimmed wagon. The narrow rims reduced friction on the dry, but hilly road from the wharf at Penticton to Nickel Plate.

Herb Clark Estate photo

result of Clark's many years of experience as a surveyor and railroad grade contractor.

The Green Mountain Road follows much of the original freight route. The switchback sections of the road leading up to the mine and the Nickel Plate townsite were used during the mid 1930's to transport mine machinery to the mines. Clark's roadhouse was in an ideal location for both freighters and the stage line. Freighters made this an overnight stop on their way to the Nickel Plate Mine. They could obtain food and lodging for themselves, and hay and a barn for their horses.

The Welby Stage Line, and later the Tweddle Stage, made this their noon stop, arriving about one p.m., after leaving Penticton early in the morning. The stages changed teams here and left to make Keremeos Centre by nightfall. The Green Mountain Stopping House was the second stop of a five day round trip for the freight wagons to the Nickel Plate Mine. Some of the teamsters who would ply this route would be the Bassett brothers, Fred, Dick and Top, the Gillespies, a father and son team, the Brents, Phelps, Gerry Clark, L.A. Clark's son and Dave Innis, who later married Winnie, Clark's youngest daughter.

The third overnight stop for freighters was spent at the Russell House, a two-storey log structure located halfway up Apex Mountain. The story of the house is that there was at the time a very posh hotel in Ottawa, called the Russell House and it is possible the

The men in the foreground are Kidston, a Naramata resident, and Mayor E. Foley-Bennett. In the background are two "long line skinners", who have backed their teams close to the wharf to unload gold concentrate, which will later be moved out by sternwheeler to the rail head at Okanagan Landing. The men will possibly load supplies and machinery destined for a mine in the area.

Joe Harris photo collection

same name was given to the freighter's stop as a spoof, because of the rough conditions at the overnight stop. The other story is that the name Russell House came from the fact that one had to "rustle" for themselves.

Today, motorists and skiers can make the journey to Apex Mountain in less than one hour. Ninety years ago freighters had to spend three and a half days to reach this point, and could look forward to a day of difficult road before they reached the mine by nightfall. The trip down, loaded with ore, took two days. On the steeper switchback sections, wagons and sleighs were held back by chains locking the wheels of the wagons, or tied under the runners of the sleighs. The remuneration for all this was nine dollars per ton of ore coming down from the mine, and twenty dollars per ton of freight going up; plus of course, the status of being a freighter!

The status of being a freighter must have had many drawbacks. Life around the year 1900 was difficult, but the life of a freighter seems to have been a little more difficult. Most freighters spent a good part of preparing for a trip by loading heavy machinery and mine equipment into a high freight wagon with few, if any, mechanical lifters. Most heavy equipment was handled with a block and tackle, skidded or pulled up a ramp, or just plain lifted. Everything had to be loaded to ensure an early start to avoid being

These are the Gillespie freight teams hauling gold ore concentrate from the Daly Reduction plant prior to 1909 and the coming of the railroad. The freighters often travelled together to assist each other through boggy spots such as Brushy Bottom, now the George Lawrence Ranch, between Hedley and Keremeos.   Joe Harris photo collection

caught on the road at night, or to allow some extra time for misadventure, such as a loose rim on the wheel or broken harness.

In the morning the horses and teamster had to be fed early. There were over eighty 4 and 6-horse teams freighting out of Penticton in its heyday, not all would use a barn or a livery stable. Those horses that stayed out were tied, usually to a wagon wheel, and depending on the weather, covered with a blanket. The driver usually slept under the wagon. Horses must be curried or groomed, checked for sweat scalds, harnessed, and hitched to the freight wagon. All horses were not wildly pleased about this procedure, and some would resist in annoying or downright defiant ways.

Once the outfit was underway the driver could relax by watching for rocks on the road, or chuckholes which could cause a load shift or a breakdown. The wagons used by most of the freighters had very high narrow wheels, the reason for this was that they cut down the resistance and made the load lighter. The narrow wheels did create problems at the sandhill behind the Indian Church near Penticton. It had to be climbed to get onto the original freight road. According to Sandy Brent, whose father drove a freight outfit, the teamsters often doubled their teams at this spot and pulled each wagon up with twice the amount of horsepower. The freighters then camped on a flat up above the Indian village near a stream. Here there was a log barn and some accommodation for the teamster. Sandy Brent also recounted a story of one large tank that was being taken to the

Front Street, Penticton, 1907. A cut of the cards determined that Bill Garrison owned the livery part of a hotel and livery business owned jointly by Bloomfield and Garrison in Midway. Garrison moved the business to Princeton in 1907, and later contracted to haul two steam boilers complete with smokestacks from the Penticton wharf to the Nickel Plate Mine. The boilers had to be unloaded from the boat and set on wagons, after which Joe Wilson and Frank Garrison drove the six horse teams, hauling the boilers. Bill Garrison handled the team with the smokestack and extra supplies. This photo was taken on Front Street, across from what is now the B.C. Motel.

Ernie Garrison photo

mine by his father. Twelve horses were required to pull the wagon loaded with the large tank, a steam boiler for a steam compressor, that was being taken to the mine up the sandhill. A four horse team was able to deliver the tank safely to the mine from that point.

There were two early wagon routes to Hedley, one was the Green Mountain Road, and then up the tortuous switchbacks through what is now Apex Alpine ski area. The other was to continue on the Green Mountain Road to Keremeos, and then to Hedley. This was a relatively problem-free route with the exception of Brushy Bottom, a low swampy area between Keremeos and Hedley where the George Lawrence ranch is now located. The teamsters apparently doubled up their teams to haul the freight wagons through. Another route led up the Twenty Mile Creek canyon crossing the creek on crude bridges twelve times in three miles, then continuing up through the Golden Zone and then on to the mine site at Nickel Plate. This route seems to have been used more for transporting goods between Hedley and the Nickel Plate Mine. The route had the disadvantage of not only the twelve bridges and a very steep grade, but a swamp area near the summit.

Joe Wilson and Frank Garrison stopped their teams on Main Street, Penticton, to pose again for the cameras. The wooden sidewalks in Penticton, which at this time was incorporated as a municipality, protected pedestrians from mud, or dust of the unpaved streets. A.H. Wade's store in the background of this photo is on Front Street. When it burned, the property became a gas station, and is presently the Muntz Stereo Centre. Joe and Frank probably had to double up their teams to haul the boilers up the sandhill on the Indian Reserve before reaching the plateau where the road was located. They would camp there for the night, under their wagons or at a barn that was located on the Sandhill road.

Ernie Garrison photo

Ernie Garrison lives in Princeton. He talked about his father who was in the horse-freighting business and later the trucking business.

"My dad was Bill Garrison. He came from Ellensburg in the States 1898-99. The first job he had was driving stage from Okanagan Falls to Camp McKinney. He drove stage for Gillespie. He went to Midway and started up a business with Mr. Bloomfield. They decided to move to Princeton. They owned the hotel and they also owned the livery business. The men cut the cards to see who would own things. My dad won the livery business and Bloomfield won the hotel. Bloomfield ran the hotel in Westbridge until it went belly-up. He went out of business and came to Princeton and bought a half interest in the Princeton Hotel. At that time it was called the Great Northern Hotel. It burned down later.

We came to Princeton in 1907, so it must have been 1908 or 1909 when these pictures were taken in Penticton. My dad went from Princeton to Penticton to move these boilers up to the Nickel Plate. There were two 6-horse teams and one 4-horse team on Front Street in Penticton. The four horse team would be my dad, Bill. My uncle, Frank

Garrison teams at Nickel Plate. Hauling the boilers on the Green Mountain Road, which was then located on the dry plateau above the creek, on to the Green Mountain Stopping House was relatively easy. The problem began when the six horse teams and heavy wagons tried to negotiate the multitude of switchbacks on the steeper sections of the Old Nickel Plate Road. The Teamsters had to block the wagons, unhitch the teams and use a set of blocks and a line to pull the wagons around the tight curves. It took the teamsters five days to get the boilers from the bottom of the mountain to the minesite on Nickel Plate Mountain!   Ernie Garrison photo

Garrison and Joe Wilson, had the boilers. Dad had the smokestack and all the rigging leaving Penticton. They loaded the boilers off the boat onto the wagons down there at the dock. The boat had come down from the rail head at Sicamous, where the freight had arrived by C.P.R.

It took them five days to get from the bottom of Green Mountain, up to the mine. When they were going up, the road was so crooked that only two horses could pull the load, and they couldn't manage, so they used a block pulley and line to get around the curves. The teams would only pull the wagons off the road, so they used block and line to get them up the hill.

My dad had about 46 head of horses in the barn in Princeton. He had three teams going up to Copper Mountain and three teams coming down. The team that came down would load up the next day to go back up the mountain. They hauled water, machinery and everything up to Copper Mountain when it first started.

They hauled all the steel for the railroad. They didn't have any bridges built in the railroad, so they cut the steel, then loaded and hauled it down to where the bridges were to be built.

They also hauled coal from Blakeburn to Coalmont. The teams that stayed up in

Elwood Bromley, shown here with Dave Innis' team in 1912, started work as a teamster at a very young age. Dave Innis bred and drove gray Percherons used here as the wheel team. He also used celluloid rings as decorations on the bridles of his lead team. The wide wheels on the wagon and the freight bells on the off leader indicate Elwood was doing some serious freight hauling.

Bud Gawne photo collection

Coalmont didn't come down to Princeton. It took them all day to go from Coalmont to Blakeburn, five miles. It was all uphill. They loaded up the wagon or sleighs, or whatever they were hooked up to. They had to shovel that five ton of coal off into a railroad gondola car at the railroad tracks in Coalmont. From the gondola it came down to the C.P.R. coal stations on the railroad.

Quite a few of my father's horses were Clydesdales and Percherons and good horses of mixed breeds. He put housings over the collars, so that when it rained, they wouldn't get wet. All his horses had housings on them. He thought a lot of his horses. He really looked after them. He took blankets out to them after a day's work in the wintertime to cover them up."

Bud Gawne recalled a conversation with an old time freighter, Elwood Bromley. "They'd have to double up on that sandhill, past the church. He'd take a team down and hook on. He was only 12 years old, and he'd ride the team of horses. That's when he started with Innis'. I got a picture of him on a freight, driving. He'd drive and he was just a young fellow. He told me that was a day's trip from Penticton up over the sand hill to where they had this horse barn and camp. I would like to take you and show you the whole setup of this. You can still find some of the road.

Another thing Elwood told me - they were all good drinking men - and he said every guy carried his own bottle, that was on his wagon by his seat, because the whiskey was sold in bulk, they'd bring it in in barrels. He said for fifty cents you'd get a 26 oz. bottle. It was mostly Irish whiskey and stuff like that. He wasn't a drinking man. I asked him if he ever saw any bottles. He said, "I've seen lots of them there." He drew me a map, he told me where to head up. He said, "You follow that trail, because there's a good sidehill along there. When you come to a flat, look down and you'll see Shingle Creek." It was all up on top on that side of the Creek. The first bottle we found my wife Kitty picked up on top of the gulch.

He'd get back with his horses and maybe make four or five trips a day, and he said most of the outfits in those days were six up, so it took eight horses to pull a wagon up that sandhill! That was a day's work. Then the next day they'd take off. They had a barn, a bunkhouse, and a cookhouse.

The barn and camp lasted quite a few years I guess before they got that road that followed along by Shingle Creek. It never crossed Shingle creek at all, it was on the other side. I followed that and you can still see it. Sometimes there'd be 20 wagons coming through there. They had a cookhouse and bunkhouse. It was a narrow son-of-a-gun, you could just get across there with a team and wagon. You know how steep that is from the church to that sand hill. Elwood Bromley said sometimes he'd ride his horses back and forth up to five times a day. He worked for the freighters for years. He worked driving teams when they built the Great Northern Railroad grade from the border to Hedley."

J.A. Nesbit in The Penticton Herald, December 23, 1911, under the heading "Recollections of an Old Timer" had these reminiscences of the early freighters:

"Those were the days of freighting. Previous to the building of the V., V. & E. to Keremeos, all the freight for points on Similkameen, and south, was handled here. The town was alive with freighters and freight teams, 4 and 6-horse, pack trains intermingling on the dock, loading. The night was lurid with camp fires, loaded wagons were placed where possible to leave road space, ready to pull out in the morning. The hotel did a roaring trade. An occasional yell at night didn't alarm the residents, they were used to it. But the advent of the railroad has stopped that industry. There was money in the freighting business then, all the heavy machinery for the Nickel Plate mine at Hedley, passed through here. It is wonderful how such heavy castings, weighing from 50 to 60 hundred, were loaded on the wagons, and how they managed to haul such loads over the rough roads we had then over the mountains, and reach their destination in safety. A large, heavy cable for the tramway from the mine to the town in Hedley, weighing several tons in one piece, had to be loaded on three wagons. One part of the coil was on each wagon with the cable trailing on the ground on one side, from wagon to wagon. These teams had to keep the same distance apart all the way, and roundly cursing all obstacles in the way of rocks and stumps, they got to Hedley all right."

6. EARLY HEDLEY TOWNSITE

Camp Hedley at the base of Nickel Plate Mountain was originally part of the land set aside as Indian Reserve. Frank Bailey, a mining engineer from the Boundary country, noting this, purchased a large tract of land from the provincial government near Boulder Creek, a tributary of the Similkameen, upstream from Hedley and Twenty Mile Creek. Bailey had R.H. Parkinson survey a townsite, which was to be called Similkameen City. M.K. Rodgers apparently appealed to the government also, and had the Indian land resurveyed and two fractions removed from the Indian allotment for the present townsite of what was to be Hedley City. The Indians of Chuchuwaya near the Oblate Mission Church, were upset about the Indian land being annexed by the new development. Their chief medicine man, Cosatasket, invoked the spirits and put on a curse, presumably on the land. The curse apparently didn't work. Rodgers later acquired land from the government for a slag pond in connection with the mill that was to be built.

Sometime in the autumn of 1899, Thomas Bradshaw purchased a recently built log stopping house from a man named Johnson. The building was located at Fifteen Mile Creek, later named Bradshaw Creek, between Keremeos and Hedley Camp. The hotel became known for warm hospitality, good meals, and a suitable bed. The Bradshaws planted an orchard, started a dairy herd, and raised beef cattle. The Bradshaw Stopping house has long since been

Shatford's Store, September, 1903. F.H. French, store manager is standing in the doorway. George Cahill is on the right. E.D. Bowing, with hands in his pocket, is at the corner of the building. Myrtle McLean is the girl on the left.
Pat Wright photo collection

J. A. Schubert, General Merchandise, September, 1903. Note the Twenty Mile Creek bridge and the beginnings of the mill above the store. Schubert started a store in Penticton at Vancouver Ave. and Ellis Street, moved to Hedley, and eventually relocated to the settlement of Tulameen near Otter Lake, in the Tulameen Valley.   Pat Wright photo collection

The first log bridge over Twenty Mile Creek in early Hedley. Sandy Brent, Eddie Brent, and Roy Jakins later reconstructed this bridge. Sandy recalled, "We drove piles with an old horse and a little A frame. The horse, owned by Jimmy Jamieson, would pull the hammer up and there was a trip rope on the single tree to let the pile driver fall.   Pat Wright photo collection

In 1903 L.W. Shatford put up a one-storey building on Scott Avenue and opened a general store with F.H. French as manager. Two or three years later the store building was greatly enlarged and the firm's name changed to Shatford's Limited. The old building still stands, and was occupied by Collen's Department Store. Herb Clark Estate photo

torn down. Bradshaw Siding, a name given to a railroad siding, canyon, and creek near the old hotel is the only indication of the original stopping house. The railroad has since been removed also. Tom Bradshaw did have some mineral claims above the Mascot Mill.

In later years the Bradshaw and Lawrence ranches were close enough together that the ranch owners could help each other on certain ranch activities. In 1946 George Lawrence had trailed a mixed herd of Bradshaw and Lawrence cattle to Princeton, to be shipped out on the train to Vancouver. George visited Lenard Bradshaw who was in the hospital in Princeton. Lenard later died. The day before he was admitted to the hospital Lenard had shown George Lawrence a cheque from the Hedley Mascot Gold Mine for $42,000. To George's knowledge the cheque was never taken to a bank by Lenard before he died.

Carl Loomer, who was a long time employee of the mill and resident of Hedley had this to say. "The Shatford store was about the first building in town, the first real store. Schubert's was ahead of it, but Schubert's was a very small affair. Everything was in Shatfords. The store that you see on the opposite side of the street from it now was built by a man named Finley Fraser. They put the second storey on top of Shatford's store, and that was the Masonic Temple and dance hall. We danced all night. We had eight by eight posts to make the store secure, but still everything fell off the shelves and all the glassware broke. So we had to quit dancing, it was considered a thing of evil.

Hedley circa 1910. The Great Northern Railroad has built a station and there are box cars on the track, (photo, top right). The mill has started one slag pond, however, there are still construction materials stacked nearby. In the town of Hedley, the business area has built up with stores and hotels.   Doug Cox photo collection

They started production at the mill in late 1904. It was a very small mill when it started in production. The last foundations that you see over there now, are twice the size of the original mill when it first started. The railroad came in here in 1909, and there were no more four horse teams or anything after that.

In the early 1900's the building boom of Hedley took place. The first merchants were J.A. Schubert, and W.T. Shatford & Co., both General Merchandisers. The Shatford store was opened in Hedley May 1903 by Frank French."

Frank French was born in England in 1875, and came to Canada as a young boy. After gaining some experience in the grocery business, he went to Fairview in 1897 to work for Shatford. In 1908 he married Anna Beatrice Brown, who had come to Hedley to visit her sister. Mr. and Mrs. F. French bought Shatford out in 1913, and ran the store for five years, before moving to Vancouver. Frank and his wife returned to Hedley in 1936 to work for the Innis garage, he later worked for the Mascot Co. before he retired in 1950.

Reginald H. French was born in Hedley in 1915, in what is now called the Mascot guest house. He moved to Vancouver in 1918, then to Vernon in 1920, and back and forth between Hedley and Vernon in 1934, 35 and 36. He worked at Sterling Creek Mine in 1935 for three months, and Gold Mountain Mine in 1936 for three of four months. He started work on construction of Mascot Mill about 1939. After construction he worked in a carpenter's shop until

This building was once Burr's blacksmith shop. Carl Loomer, who grew up in Hedley and later worked at the mine, recalled running to the hotel for a lard bucket of beer for Mr. Burr. He was given 10 cents to pay for the beer, however, Burr had a tab at the hotel so Carl got to keep the 10 cents.
Doug Cox photo

1943, and later as an operator in a Cyanide Plant. He left Mascot later in the year and he and John Lawrence purchased Valley Garage from Norman Armitage. He later bought Lawrence out in 1946 and took George Pizzi in as a partner in 1947 for a couple of years.

Reg recalled his working days in Hedley. "Besides operating the service station,our main business was trucking, hauling coal, wood, fuel oil, sawdust and general trucking, as well as hauling furniture or freight to anywhere in B.C. After getting rid of the garage I went to work constructing buildings at the Oregon Claim, and after completion, I operated the scraper until freeze-up."

French left Hedley in the fall of 1952. "I was member of the Moose Lodge all during my stay in Hedley. I was married in May of 1944, and the Lodge put on quite a party for us on our return from our honeymoon."

Carl Loomer remembers E.E. Burr, another resident, very well.

"He was a blacksmith in town when I was a kid. He used to give me ten cents to go to the bar and get him a lard can full of beer. His credit was so good I didn't have to pay for the beer, they'd charge it to him. When I'd get back to the shop he would give me a dime. I don't know whether it was a five pound or three pound lard pail." Carl went on; "He made a lot of money in here in the days of the horse racers, he was a very outstanding blacksmith, and he could tell pretty well whether a horse was going to be any good or not, and usually he did very well."

The Hedley hospital was incorporated, and in the spring of 1907 the new building was erected by Messrs. Boeing and Brass. "The building is a three storey structure 24 x 40 feet, with a wing 16 x 26. The lowest floor or basement contains kitchen and laundry, and the main floor, entrance to which from the outside is made from the hillside, comprises the hallway, one large five-bed ward, two private wards, operating room, and bathroom. The third-storey is yet unfinished, but could supply two comfortable bedrooms for nurses." It opened in 1910. The first operation was performed in the hospital on February 23, 1910. Dr. M.D. McEwen was the surgeon-in-chief, and Miss Bond and Miss Fraser, both of Vancouver, were the first two nurses. The hospital was closed down in the fall of 1930 and never reopened.   Doug Cox photo collection

Carl chuckled; "He knew which way it was going to go after he put the shoe on."

Ambrose Tree is recalled by some as one of the more colorful characters in Hedley. He was one of those fellows who would do anything for anyone, and not expect much remuneration for his trouble. At one time, he hauled coal from Princeton in an old Model T with a wooden box which had a dump mechanism. He stopped and talked and visited on his way from Princeton, so it was usually late by the time he stopped at Mrs. Kate Bromley's for a meal, before going on to Hedley.

The lights on model T's, as some of us will recall, were run off the generator. Since early Fords had a magneto, no batteries were necessary. This system created one perplexing problem; the faster you went the more light you had, if you slowed for a bump or something similar on the road the lights dimmed almost out. Ambrose Tree solved this annoying habit by hanging a gasoline lantern in the front of his truck to supply a constant light!

When electric welders first became available, Ambrose obtained one. He did custom welding for people at remarkably reasonable rates. He was also a mechanic, and worked on cars until he couldn't stay awake any longer. The front of his shop had an old mattress on the floor,

In the winter of 1900 D.G. Hackney built the Hedley Hotel on Haynes Street, a neat two-storey hewed-log building which was opened early in March, 1901. It burned down as did many of the hotel-type buildings.
Doug Cox photo collection

where he retired until he was suitably refreshed enough to start work again. Ambroses' coveralls being very greasy, deposited some of this on the mattress, which was apparently black with grease and grime.

A man with such peculiar work habits is sure to have other peculiar habits. Ambrose's other eccentricity was eating. A sandwich was a loaf of bread cut in half lengthwise and filled with the contents of one or two cans of sardines. On one occasion he went to Mrs. Mauleys, and, being rather hungry, stowed away two dozen hard boiled eggs at one time.

Carl Loomer recalled. "The year the mill started, M.K. Rodgers was the superintendent at the time, there was an accident. The first here, and it was a death. Charlie Allison's father, gave a piece of ground to the people of Hedley for burial purposes. They took their graveyard out and took it down to where the big white church is. Charlie Allisons' father was an Indian. They didn't take their people out; they just abandoned their graves. In 1918, because of the influenza, they put 22 graves in the new site. There are five graves in another graveyard on the other side of the mountain on the original Nickel Plate Road leading to the mine. Two of them were miners that were killed underground and the other three were babies, from families that lived up at the mine. They lost their babies, not from the flu. One in particular was Mrs. Dave McLelland, her first one. That's on the right hand side as you go to the lake. At that time we called it Strayhorse Lake. It's now Clearwater Lake. It was also called Nickel Plate Lake."

Once Sandy Brent came off a shift at the mine with measles. He wouldn't be let near the

In the summer of 1904 the Similkameen Hotel was built and opened for business in the autumn. It was a modern, well-built, and comfortable hotel and soon became a popular stopping-place for travellers. It was burned down in February, 1916.   Doug Cox photo collection

bunkhouse so he walked down the mountain and went to Winklers. There was no place to stay at Winklers so he went to the hospital. Dr. Wride put him in quarantine under isolation into the unfinished room at the top of the hospital for two weeks. Bush rats ran around the upper roof portion. The nurses left food for him at the bottom of the stairs. Finally, Cliff Clark came to visit so they sneaked out and went out fishing up Twenty Mile Creek during the day.

Mrs. Myrtle Hardman was born, married, and raised her family in Hedley. She came from a family whose ancestors date back to the early history of the area. Her grandfather was John Bromley and her mother Eliza, married Mr. Jack Edmonds. Jack came to the Okanagan with a carload of prize bulls consigned to Tom Ellis, the land baron of the Okanagan. Myrtle married John Hardman. She and her husband had both grown up in Hedley.

Myrtle and her sister, Ivy, and their parents lived in the family butcher shop. For those of us who can only remember purchasing meat neatly packaged in cellophane, beautifully displayed in a supermarket cooler, presided over by white jacketed attendants, an early butcher shop would be an experience.

Mr. Edmonds was a butcher. He bought his meat on the hoof from local ranchers, then butchered and dressed them in his own slaughterhouse. Myrtle, his daughter, remembers that the slaughterhouse, which was located down by the river, had a chute with a catwalk from where her father would shoot the animal in the back of the head. After the animal

Charles Richter of Keremeos, built a two-storey building as a butcher's shop and residence, and he supplied the town with its meat until he sold out the business to Cawston and Edmond. Shortly after this, John Meyerhoffer became associated with the business, first as store manager and later as owner. His connection with the business continued until 1931, when he sold out to Eugene Quaedvlieg. Edmonds had his slaughter facility located near the river. This was the hand cranked winch which lifted the beef carcass so the beef could be quartered. The beef, if not frozen, was kept in a building cooled by ice.   Doug Cox photo

dropped he would go down and bleed it. Jack Edmonds had the necessary equipment to haul the animal up where it was gutted and the entrails shoved out a trap door. There was a large pit nearby where Mr. Edmonds laid the hides layer on layer, hide and salt. The carcasses were hung in quarters on a large rack where the meat stayed all winter.

"No one, but no one, touched any of that meat!" exclaimed Myrtle.

When the butcher wanted more meat for his shop he brought in a quarter, which was frozen, and hung it in the cooler. When it thawed it was ready for sale. The store didn't have a fancy name, it was Edmond's Butcher Shop. It was the only butcher's shop in town.

To keep the meat fresh in the summer, it was kept on ice from the ice house. The walls of the icehouse were packed with sawdust for insulation. Blocks of ice cut from the river in the winter were stacked in sawdust with sawdust packed around each block of ice to prevent it from melting too quickly. To add additional ice time to the life of the ice in the icehouse, Mr. Edmonds started his cold storage area by flooding the floor with water which would freeze, giving a good base to the ice which was going to be stacked. Myrtle and her sister Ivy thought themselves very exclusive when they could use the frozen base of the icehouse as their own private skating rink!

In 1903 Dr. J. Rolls opened a drug store and office. It was later
operated by his son Lou Rolls.   Rolls Family photo

Ice was a commodity in the early 1900's, in the days of iceboxes and before mechanical
refrigeration. Ice trains used to pass through Hedley bound for the packing houses and cool-
ers somewhere in the United States. Those trains got their cargo from Otter Lake near
Tulameen. There was a substantial ice-cutting industry to keep trains supplied with what
would be a very seasonal product.

Margaret L. Stein, presently a resident of Newport Beach in California, was born in
Hedley in the year 1915. She recalled that some of the best times she can remember were
while she was in Hedley. Her father, Louis C. Rolls arrived in Hedley about 1904 to manage
the store belonging to his uncle Dr. James Rolls. In 1911, Louis had his brother Bruce manage the store for him while he went back to Detroit to marry Margaret's mother Annetta.
Louis and Annetta lived in Vancouver for a year, where Wallace Rolls was born, and then
came to Hedley.

The graveyard, located west of Hedley, was donated to the citizens by Charlie Allison. Twenty-two graves were added when the influenza virus spread at the Nickel Plate townsite. The bodies were brought down the tramline to Hedley and buried in this graveyard.   Doug Cox photo

Mrs. Stein forwarded a postcard showing a view of the Daly Reduction Co's Stamp Mill and Cyanide Plant. The publishing date was 1908. The store was sold sometime in the early 1940's to Maurice E. Winters, and it became the Red and White Store.

Louis Rolls passed away in Princeton in 1959 at the age of 88. His wife Annetta, passed away in Vancouver in 1968 at the age of 91. Bruce Rolls and his wife Effie and their family Warren, Katherine, Doug and Dorothy lived in Hedley for quite a few years.

Margaret Stein (nee Rolls) recalled some of her experiences and memories as a child, youth, and young person, in the community of Hedley:

"I started school at what was the Old School, located underneath the mountain. There were slides there and the area was condemned. The New School was built near the road on the Princeton side. I never attended that school myself. I do remember that they once presented me with a large book titled "Chatterbox", signifying that it was meant for me. Also, one of the teachers was forever after Fred Corrigan and myself for talking in school. I do recall at one time I had a Japanese orange box under my feet because I was so small, my feet wouldn't reach the floor. Another recollection I remember, is a teacher, Ralph Thomas, who used to get pretty discouraged when a lot of the pupils chatted all the time. He would wind up and throw chalk at them, to see if he could score a bullseye!

Most of our best times were during summer vacation. We trekked about a mile and a half or more to the old swimming hole that was above the dam. There was a strong current in the

Members of a convention in front of Shier's Clothing Store in Hedley. There was a large Masonic Lodge in Hedley, started in 1906. Perhaps this was a Masonic Convention.

Herb Clark Estate photo

river. I could hardly swim across. Sometimes we'd swim over and walk up the other side of the river and find big logs, and come back down with the current on top of these logs. Pretty risky when I look back at it now. We had great times down there. The water was cold, but very refreshing. We often made this walk twice a day, afternoon and evening.

In winter, skating was the big thing. We used to have the rink up by the old school, just a little shack for a change room. We had some wonderful times. Later on when the school was moved, the rink was near the mill and the powerhouse. I can remember the nights when it was so cold it was a wonder the shack didn't burn down. We had a little potbellied stove that would be crimson from overheating; it's a wonder we survived at all.

One time my cousin, Doug Rolls, and a lot of the boys borrowed the company firehose, and used it to flood the rink. It was a really cold night and the boys apparently left the hose up at the rink where it froze solid. It would have been a bad state of affairs if there had been a fire.

We had lots of good dances in Hedley. We used to have orchestras sometimes brought down from Princeton. We'd dance until 2:00 a.m. About then no one would want to quit, so we would take up a collection and pay the orchestra to play another hour, and we would go home hardly able to walk with sore feet.

One of those dances was held the night there was a big slide up above all the Mascot company houses, just below the mountain. We didn't know a thing about it because of the noise

Hedley golf links Christmas Day, 1919.   Pat Wright photo collection

of the music. Later we were up in the cafe having something to eat. It was so crowded, some of us were in the kitchen helping the Chinese cook make sandwiches. After I got home and was in bed we heard the fire bell, got up and discovered that a terrible slide had taken place with huge boulders bouncing down. There were only two people killed, but there was quite a bit of damage to some of the houses. The houses were eventually moved into the main part of town.

Some of the things that happened were pretty frightening, I can remember once, some years before the slide, playing in a tent on a neighbour's lawn and we heard the fire bell. That I believe was when the big fire took place and devastated the hotel that was behind our house.

I can remember very well being in the tent and being frightened but I can't remember the fire itself. I also remember the fire at the top of the mill. We watched that fire from the window of our house and I remember how terrifying it was.

At one time, when I was working in the store, we started a group called, 'Our Happy Gang'. It was for any single person in town or any person who came to town, age 19 and up. We started out with about fifty people and we had some really good times. We set up groups that were in charge of entertainment for the month. We rented the community hall and had special parties. In our group we even had an orchestra, Thelma Gordon played the piano, Fred Gordon, (no relation), played the trombone, George Angus played the violin and sax, and Frank Lines, the policeman, played the drums. We had the basis for some good times.

I remember one New Year's a few of us drove up the old back road to the Nickel Plate,

Sports teams were popular in early Hedley as seen by this Football Club. Football at that time was the game we call soccer today. The team members are: Top Row: Will Richard, Lou Barlow, Harry Pilkington, Charlie Hambly, Scottie MacLaren, Dick Hambly. Middle Row: George Frances, Bill Graham, unidentified. Bottom Row: Jack Pilkington, Bill Frances, unknown, Charlie Luxon, Jack Trudgeon.   Doug Cox photo collection

where we visited friends and went to the dance. It was an absolute fairyland up there with the snow. In Hedley, it was usually muddy and dirty looking. We always went up to the mine and got our Christmas tree.

I remember that we had Mr. Butler, a negro barber, a really nice man, well liked by everybody, who came down on an ore car from the mine. The ore car came very fast. I don't know whether something really happened or if it was just usually that fast. The men at the bottom said it was sure a white black man that came down that day!

I was thinking back to our days of sleigh riding. We used to use the old hospital hill, which at that time seemed terribly steep. We came riding down over the bridge above the creek. I remember Herb Neil coming down and crashing into the bridge. He had the most terrible cut on his head. I can still visualize the stitches to this day.

My Dad raised canaries. He used to ship them to every state in the United States, parts of Canada and the Northwest Territories. He used to put notes in poem form on all of the shipping cages as if the birds were asking to be fed and watered while on route. We had a bird room right off the living room. When I was 10 years old I had to have my tonsils and adenoids removed. At that time the hospital up on the hill was much too cold, it was in the middle of winter, so the operation to remove my tonsils and adenoids was performed on the dining

Hedley, June 12, 1914. The riders are from L-R: Billy Gibson (later Mrs. Lillian Estabrooks), Billy Daly, Kay Gibson (Billy's sister), and Florence Daly.   Estabrooks photo

room table in the living room. After the doctor was all through he suddenly remembered about the anaesthetic and said, "Oh Lou, I forgot all about the damn birds." As it was they lived to sing again!

I was thinking back to the days when the mill was closed down for the winter and I used to take the census for my own amusement. I knew everyone in town and at the mine, including kids, in winter they numbered about 300, in the summer there were about 800. Of course later I didn't know them all.

I remember breaking the ice on Twenty Mile Creek to get buckets of water to bring back on our sleighs. I was thinking back to when the power was off and we went from coal oil lamps to the new gas lamps. They were great, so bright, but scary to light."

Howard Graham while in hospital at Princeton, shaved Jack Budd, supposedly Bill Miner's partner. At that time Budd told Howard, "You tell your boy you shaved a man over a hundred years old." Jack Budd used to have a ranch out of Princeton, and the nurses would ask him to tell them about Bill Miner, but he would clam right up. There was a story going around about him having money hidden out in the hills, but Jack wouldn't say, Boo."

Howard recounted that no one could get any information out of Jack Budd,

"Cagey old son of a gun, good eyesight! I asked him for the time one evening ; it was dusk, hard to see, but he glanced at the big gold watch on the table and said, "Eight o'clock Boy!" The nurses all had safety razors and I had my straight razor. I didn't have my strop so it wasn't the sharpest, and old Jack was a tough customer to shave, but he begged me to shave him,

George Cahill's cabin in Hedley. Walter Jamieson lived in the cabin in later years. George led the first pack from the Fairview train over the Camp Rest Trail with supplies for the Nickel Plate in 1898.
Doug Cox photo

so I did it. When I finished he told me that I was the only man with a razor, the rest of them people just carry a hoe!"

Howard went on to describe Jack Budd. "He must have been a tall man in his youth, but in the later days, I can see him yet, going along on his horse through Princeton, dark clothing on, humped over his horse. He must have been a big man when he was young."

Mrs. Myrtle Hardman whose grandfather was John Hatten Bromley, for whom Bromley Rock is named, recounted what her mother had told her about Bill Miner, alias George Edwards. "Jack Budd had a ranch outside of Princeton. His house was built right out in the middle of an open field. My mother, then Eliza Bromley, could never figure out why because there was a grove of trees not far away. My mother was a cowgirl, when she was in that area checking on cattle, George Edwards and Jack Budd would ask her in. They usually put on some sourdough pancakes and fried some pork; Edwards always had to say grace as soon as they sat down at the table.

~ 62 ~

Branding in Hedley. Hedley may have been the mining boom town of the Similkameen, but the Similkameen was still cattle country. The cattle indicated shorthorn breeding, before the introduction of Herefords. The flat roofed building is thought to be the Grand Union Hotel.

Doug Cox photo collection

My mother had always thought, after she found out that Edwards was Bill Miner, that Jack Budd had taken the horses from the ill-fated train robbery at Ducks and that he possibly got the loot. After Bill Miner and his gang got caught, Jack Budd went into race horses and he had some good stock. His nephew, Clyde Wilkinson, who used to be a butcher in Hedley and then moved to California, came up and bought some of these horses from old Jack and took them down to the States to be used as polo ponies. Now, where did Jack Budd get the money?", was the question Mrs. Edmonds asked.

"Mom, Eliza Bromley, was just a kid when she knew Bill Miner. He always claimed that he had a silver mine down across the line, any time he went broke he went down across the border and always came back with money. I guess he went down and robbed a stage or a train or something. The one and only thousand dollar bill my mother ever saw was shown to her by Bill Miner. He had a money belt and took out the bill to show to her.

Bill Miner always wanted to know what my mother wanted brought back from these trips and she usually asked for candy or something inexpensive. One time he brought her back a pair of spurs. She always said they were silver spurs, however they didn't last very long and someone got them. I wish I had them now!" was Myrtle's comment.

Ralph Overton, now in his nineties, lives at Williams Lake. His family moved to Fairview in 1920. He has a wealth of experiences to share regarding the time he spent in the Okanagan

These were the Hedley recruits August 24, 1915, who had signed up to fight in the Great War 1914--1918.
Doug Cox photo collection

and Similkameen. He did work on the building of the Mascot Mill in 1934, but in 1930 Ralph worked for the telephone company. "We started in Osoyoos and built a line over to Hedley and Princeton. We camped in Hedley one winter and boarded with Mrs. Bromley. She was Bill Miner's girlfriend at one time. Eliza Bromley was the mother, Myrtle Edmonds was her daughter. She used to talk about riding through a little canyon following a mountain trail to Jack Budd's ranch. Jack Budd's ranch was at Princeton. The Bromley ranch was between Hedley and Princeton. Mrs. Bromley, Eliza was her name, was a very pretty girl, they said. Old Bill Miner used to like to dance with her. She used to ride over and have dinner with Bill Miner and Jack Budd, and then she would ride back to the Bromley Ranch. She said she never knew they were train robbers. Mr. Edwards, a name he used as an alias, told her he had rich copper mines in the states, and many talents.

George Edwards loved to dance and he would put on dances, pay for the orchestra and everything, and people came from miles around by horse and sleigh to attend. He liked to dance with the young ladies. As I said before, the Bromley girl was a very pretty girl. She still was attractive when she was older. She married old Jack Edmonds because he had nice saddle horses. She liked horses. Jack Edmonds had a butcher shop in Hedley, and she opened a small cafe and boarding house.

Edwards and Budd always had good horses, especially old Jack Budd. One time she went over and the two horses, Nell and Fancy, were all stretched out laying in the corral dried

The Hedley Brass Band, 1914.   Doug Cox photo collection

sweat rings all over them. Eliza asked "What's happened to Nell and Fancy, they're all pooped?"

"Oh" Edwards replied, "If you had travelled as far as those horses went last night you'd be tired too!" They had been down robbing a train or something, and they would travel a long way because their horses would run well. Old Jack Budd, they claim, was a brother of Edwards. I wouldn't say for sure but in later years he was breeding and raising fast horses for the mounted police. He had a hundred head of horses, and he died when he was about 100 years old."

Ralph also described how Bill Miner, alias George Edwards, got caught after a train robbery at Ducks, near Kamloops. "They had their guns hidden and they sent Jack to get the horses. They had camped. They kept one wrangling horse, and Jack had gone for the other horses. He looked down and saw the Mounted Police by the camp. Bill said they were prospecting, and he had almost convinced them that they were. Then their one fellow started to walk away, and they asked him where he was going and he started to speed up. Then they shot him in the hip, and looked under the canvas and saw all the guns. The rest is history."

Ralph's younger brother Kenny stayed with Jack Budd one winter. Budd was apparently always afraid somebody was going to shoot him in the back. He had a big heavy chair with 2-inch planks on the back. He had a kind of hump-back but he would settle down in the chair with his six shooter. He had that right there beside him all the time, and he used to be afraid someone was coming. "You hear that noise, Ken?" "Not really, only the wind". Jack was

The Grand Union Hotel after the fire. These are the Winkler children after the fire that destroyed their home and parents' business New Year's Eve, 1918. The Winkler children are L-R: Gould, Andrew, Christine, Edith, Minnie and Margaret. Please note the sapling in the photo.

Pat Wright photo collection

always worried someone was going to sneak up on him. Jack Budd had been into a few gunfights. He rode up from Mexico, right through to Canada."

Andy Winkler (1911-1992), and his wife Mary (1913-1988) nee Brent, lived in a tent house while they were at the Nickel Plate townsite. Andy recalled, "When Mary and I were up there you couldn't build a house unless you put a tent on it. It had to be able to come down, we called them tenthouses. You put a tent on and nailed it around with boards, as long as there was tent showing, it was a temporary house. Then it could be torn down. Nobody could own property up there. The mine owned the surface rights. The mine gave the fellows permission to build wherever they wanted, but all buildings were considered to be on company property. Nobody but employees built there."

Mary described her mountain house. "The house had four small chairs, the first guests sat in chairs, the rest sat on powder boxes which were a useful commodity at the townsite. There were six families that were very friendly. The wives took turns cooking for the men when they came off shift at 1:00 a.m. There was Dick and Mardie Gougeons, Sam and Opal Meldrum, Val and Mary Pederson, Howard and Anna Powell, Bill and Claudia Morgan, and Earl and Violet Edwards. The grocery bill was about $30.00 per month.

Meyerhoffer and Edmunds supplied the beef for the area. T-bone steaks were ten cents a pound. Meat for the Mascot Mine was supplied by the Tweddles Ranch in Cawston. The

This is the same sapling tree, and Andrew Winkler (1911-1992), one of the children in the previous photo, 70 years later. Andrew married Mary Brent and for a time they lived and worked at the Nickel Plate Mine and Hedley. Andrew was also a wealth of information regarding the workings of the mine and area.   Doug Cox photo

Chinese also had a supply of meat at the cookhouse at the mine. Groceries were delivered up the skip on the tramline, as was coal or any other freight. Cliff Clark delivered furniture and the heavier items with a team and stoneboat. Cliff also hauled water for the Mascot Mine bunkhouses and cookhouses."

Mary Winkler became ill when she was pregnant while living at the Nickel Plate townsite. Dr. Gordon Wride met the skip when she went down to Hedley, and diagnosed her condition as altitude sickness. The Winklers moved to Hedley soon after that.

Andy recalled his getting married to Mary Brent who lived at Shingle Creek. They were to be married in Penticton. "Sam Orser, Theo Prest, Rollie and myself all agreed to work overtime so I could get two or three days filled up in order to take time off. We put in eight hours, drove home for eight hours, went back and put in another eight hour shift to build up three days so I could get time off to go down to Penticton and get married. You had to walk down the hill in those days. I was in a big rush to get married. So I went over to Penticton and in my rush I'd left the marriage license on the kitchen table. Bill Lowe happened to be coming over and he brought it over. Sam went up to Sicamous to see his family for the three days, Theo went on a fishing trip, and Rollie went to Vancouver. At that time you worked seven days a week, 365 days a year. No days off then."

In the evenings Andy and Mary used to go for walks as most people did because they didn't have any money and there were very few cars. At one time they were walking around and

Mary said they counted 19 bootleg establishments. The reason for these establishments was because the only liquor stores were in Keremeos and Princeton.

During the war, one needed a permit to get any alcohol. At that time if anybody wanted to buy a bottle, they had to get a permit, then you could go in and buy a bottle. Quite often people who had never had alcohol before in their life, started drinking because they could get a permit. If they couldn't have it, they thought they should. The idea in getting a permit was so that they could get an extra bottle for their friends at Christmas. Quite often people would buy these permits if they could, which meant that they could buy extra alcohol. When they bought the extra alcohol they could sell it in their homes, and these were the bootleg establishments.

The Winklers also talked about some of the sporting houses that were around. And there were four or five sporting houses in Hedley! "These houses usually had a girl and a landlord. The deal was that the landlord or landlady took half of the take.

Apparently the prices were $3 for a short stay and $15 for an overnight stay. That doesn't mean too much, except that wages for the miners were running about $4.00 a day, so it meant that the girls were making pretty good money. Fifty cents an hour was considered good pay, board and room was $1 a day, and there was nowhere to go. No one spent their money except around town.

Out of the fees, the owner of the house could keep half, and also had the opportunity to sell bootleg alcohol. The ladies were easily spotted. They were particularly well dressed as they had quite a bit of money. When they were out on the street, they didn't talk to anyone. No soliciting on the street! In fact, they were quite well behaved. There seemed to be a code among them. These girls would travel from town to town, depending where the business was most brisk. There aren't very many old hookers around, because quite a few of these girls married and became very respectable. In fact, I'm told that one of them in particular became very, very snooty after she gained a husband, a home and a family."

I pressed Andy for more information on this topic. He was not, however, about to part with any more information. His reason was that sharing the information would only cause hard feelings. He was right!

"In Princeton, there was a place called Angel Alley, which worked right up into the 1930's. There were about four or five sporting houses there. Apparently the men would go to a sporting house rather than a bootlegger. The house usually had the girls around, which was more hospitable, and one didn't necessarily have to partake of the services, but you could buy the alcohol. It was the same price as at the bootlegger's place, except that there was usually a little entertainment, somebody was singing, or somebody was dancing."

Apparently, one of the things required of the madams, was that the girls had a certificate. They were checked weekly by the doctor to make sure they weren't carrying any social diseases. This certificate was prominently displayed somewhere in the premises, much the same as a business license is today.

7. THE COMING OF THE RAILWAY

For years one of the great needs of the Similkameen Valley was a proper railroad connection to freight, raw materials and agricultural produce to outside markets. There was prolonged agitation for such a rail line and, at times, heated discussion between rivals.

The principal contenders for the rail line were the Vancouver Victoria & Eastern Railway, and the Great Northern Railroad. The Great Northern reached Keremeos in July of 1907, and in 1908 acquired the charter of the V.V.&E., and started grading a line from Keremeos to Brookmere. Steel was laid in the summer and fall of 1909. The first train from Oroville to Princeton arrived at Hedley at 11:00 a.m. on December 23, 1909. The station at Hedley was at that time a box car on a siding with steps leading up to it and shelves around the walls for luggage.

In 1934 the tracks between Hedley and Princeton were heavily damaged by high water. This section was never repaired and was abandoned in 1939. By this time, the Kettle Valley Railroad provided service to Princeton and west.

Much of Highway 3A between Hedley and Princeton follows the grade of the Great Northern Railroad. The Great Northern Railroad was forced to maintain service much longer than they had desired and with the closure of the mines in Hedley in 1955, the track was lifted to Keremeos.

The Great Northern Railroad, now called the Burlington Northern, reached Keremeos in 1907, and pushed on to Hedley. The G.N.R. used steam powered shovels to load the rock excavated from the cuts. The water wagon supplied the steam engine and the forge located in the centre was used for immediate repairs. Rev. Cameron photo

Teams and slip scapers moving in a seemingly endless circle moved the earth to create the roadbed for the railroad right-of-way. Much of Highway 3A follows the route of the original V.V.&E. right-of-way.
Rev. Cameron photo

Massive beams were used in the construction of the Great Northern Bridge over the Similkameen. The lantern is a company safety measure as cars on temporary rails were used in the construction of the trestle.
Estabrooks photo

The first passenger train reached Hedley on December 23, 1909. The first station was a box car pulled onto a siding. Arthur King, apparently a huge man, was the engineer on this train.
Doug Cox photo collection

Heavy traffic at the crossing near Hedley showing the Daly Concentrator in the background. Judging by the autos the year is about 1918 or 1919. *Doug Cox photo collection*

Bradshaw Siding, 1918, located between Keremeos and Hedley, was a water stop for the steam loco-motive which took on water from Bradshaw Creek. Coal for the trains was obtained at Coalmont, where the Great Northern and Kettle Valley Railroad shared the railroad tracks from Princeton to Coalmont.   Pat Wright photo collection

The Great Northern Railway crossed the Similkameen on a high trestle at Hedley. The dam for the hydro-electric power system can be seen in the foreground. Nothing remains of the trestle, but some of the con-crete blocks from the dam can be seen during low water on the river.   Herb Clark Estate photo

8. THE POWER DAM

The original power source for the Daly Reduction Plant, which was the Twenty Mile Creek flume and steam boilers, was found to be inadequate. It was thought that a dam on the Similkameen would solve this problem. The original location of the dam was to be farther upstream, about a mile. However, the company disclosed their plans and Shatford managed to stake the property before the company could pick up the option. The company would not pay the exorbitant, almost blackmail, price and moved downstream to the present location.

The construction of the dam took place in 1913 and 1914. The river was diverted to one side while one half of the dam construction took place. Then the contractors completed the other half of the dam. The original photos of the construction show a lot of picks, shovels and wheelbarrows around the construction site. Teams and scrapers dug the footings, teams and wagons hauled the cement and gravel. The rest was hand work. When the dam was completed in 1915, it included not only the dam, but a nine foot by seven foot flume, that led the water about two miles downstream to the power house, where the water turbine was located. The flume was pre-manufactured of tongue and groove lumber in the company carpentry shop. It was then hauled to the site by team and wagon and set up.

This photo gives some indication of the construction process in 1914. A coffer dam had been built to divert the water from the site. The cement mixer is motor driven, however, the tram car bringing in the gravel is hand pushed, bags of cement are stacked ready to be used. The Great Northern railroad trestle is in the background.

Estabrooks photo

The dam on the Similkameen was constructed to assure the complex of a more stable supply of electrical energy. This was not always the situation as the river froze in the winter, and the water for the turbine ceased to flow. Construction started in January of 1914, and the whole system was in operation a year later, January 1915. The team and wagon on the dam were part of the maintenance equipment. A stop log puller, a winch system on tracks, removed timbers to regulate the reservoir and the flow of water into the flume.   Herb Clark Estate photo

When conditions were ideal, the systems worked well. The turbine supplied power to the mine and townsites. The original system at the mill was used as a back up power plant. Across the top of the dam was a derrick like machine called a "stop log puller". By inserting heavy timbers or by removing them, it was possible to regulate the flow of water to the turbine, and all worked reasonably well. There were, however, certain drawbacks to the system. The Similkameen freezes in winter, sometimes severely, and this caused huge ice jams which had to be removed. The method used by the employees would not be considered "compensation approved" today! A favorite method was to tie sticks of dynamite to a 2x4, light the fuses and jam the dynamite as far into the ice jam as possible. Carl Loomer recounted, "A man by the name of Billy Lonsdale was doing this when something went wrong, and he was blown over with the ice and came out down below. A man was coming up from the lower plant riding a horse, their transportation in those days, and he threw a rope out there, and - by gosh - Lonsdale had the presence of mind enough to grab the rope, and they got him out! He was just like a negro for eight or nine months with the bruising he took going over the ice, and the concussion of the shock of the dynamite that went off. How he got out of it alive nobody will ever know! A man by the name of Thompson went over the dam about an hour after Lonsdale, but they never got him. They threw a rope at him, but he must have been stunned by the fall".

A 9 foot by 7 foot flume moved the water downstream from the dam on the way to the power-house and turbine. The mill is in the top center of the photo. The three mile flume was overhauled, completely relined, and some sections covered for protection against frost in 1934 by Kelowna Exploration Co. Ltd. A boiler installed at the intake transmitted steam into the water thus eliminating the freezing conditions.   Herb Clark Estate photo

If, in inclement weather, the dam wasn't creating problems, the flume did! The problem with the flume was that it iced up on the bottom, constricting the flow of water that was necessary to operate the turbine. If the turbine didn't work there was no electricity, and the mill didn't work! Cecil Jones, who now lives in Penticton, said it was nothing to see fifty men or more along the flumes breaking up ice in order to increase the flow of water. In fact Cecil can show you a knuckle on one hand which is inoperative due to having been smashed against the side of the flume while he was at work removing the ice!

Other methods were tried to remove the ice in the flume, such as adding salt to the water, but this was ineffective. At one time a large boiler was installed near the end of the dam, and they steamed the water. This worked well, however, the tremendous amount of coal used did not justify the added power. They couldn't keep up with the amount of coal needed.

One particularly severe winter, possibly 1931 or 1932, there was a power line failure near Oliver, a severe snowstorm had knocked out the West Kootenay power line, which left Keremeos, Hedley, and Princeton without power. All the electrical power resources of the Hedley area were hooked into the West Kootenay power grid, and supplied emergency electricity for the entire Similkameen. Dave Taylor, the editor of the Similkameen Star in

The powerhouse, penstock, and spillway located downstream from the dam. The water reached the forebay where most of the water was diverted into twin Francis-type turbines with a capacity of 2000 horsepower. These turbines powered an alternating current Canadian Westinghouse generator. The total project cost $192,000 - $8,000 lower than the estimated $200,000 expenditure.

Herb Clark Estate photo

Princeton, recalled that his presses had to run at night rather than during the day. By day the power was used by the Hedley Mill!

A particularly severe winter in 1935 created ice jams which caused washouts on either end of the dam. There had been washouts earlier, however, in this case the West Kootenay electrical company was supplying most of the power for the area, so the dam was never repaired! It was thought that the piers should be dynamited, so two inch holes were drilled in each cement structure, but only the center one was blown. The pillars stand in the river today, remnants of a bygone era. The flume line across the Indian land is overgrown and rotted. All that remains of the power house is a large rusted section of the penstock, few people realize that the powerhouse existed.

Carl Loomer hauled tongue and groove lumber to the flume with a team and wagon. He reminisced about the dam.

"The dam in the last years of operation was not working well because the dam overflowed. About 1915, when it first started up, the water went over the top of that stop log puller, many, many times. Eventually it went around the end of the dam, and took the building out of there. That building is seen in early photos.

Billy Lonsdale and G.P. Jones were related. G.P. was the superintendent of the mine, but

The big flood of December 30, 1917, looking down the Similkameen at an ice jam on the power dam. The employees had a unique method of removing an ice jam. They tied dynamite on a long pole and poked it into the jammed ice. The method, which would not be considered compensation approved, apparently was effective.   Pat Wright photo collection

you'd never know it. In those days he did more work than any man that was under him. They walked this flume one night and G.P.'s foot went through a plank that was on top, and he fell over the edge. He was hanging, and Lonsdale was trying to get him out of there, and all they had was his belt. He took his belt off and got him back up. I helped drill the Similkameen dam to blow it out when the mine shut down, but it was never blown, because it has twenty one-inch holes through these piers. There were four piers, if I remember rightly. They blew the center one, that's all. The rest are still there, but you can't get on it because the ends are gone, the planking and everything rotted. The Indians claim it now, so nobody argues. It is on Indian property. That dam would have been built about one mile up the Similkameen

I.L. Merrill, President   G.P. Jones, Superintendent   John D. Clarke, Treasurer

HEDLEY GOLD MINING COMPANY LIMITED
HEDLEY, BRITISH COLUMBIA
CANADA

Sept. 20, 1922

L.A. Campbell, Gen. Man.
West Kootenay Power & Light Co., Ltd.
Rossland, B.C.

Dear Sir;

Your favor of the 16th to hand. Just at present I am of the opinion that the minimum guarantee as mentioned would be greater than our people would care to assume, for we are steadily losing money, so it would not be a serious matter if we closed down for the winter. It would simply be the advisability of running through winter or closing down for the months of December, January and February.

For the balance of the year we have ample power for the present tonnage, as the 20 mile unit will provide 500 K.W. and for a short period the 800 H.P. steam auxiliary could be brought into use without excessive cost; usually January and February are good months, cold and clear. During the high water period of July we have the 20 mile auxiliary to draw on, which so far has never failed to meet all requirements. So figuring the auxiliary power consumption on the present tonnage basis it should amount to approximately $20,000.00 per year; although if installed and our development work shows the ore we expect, then the consumption would be greater.

So we respectfully request that you reconsider the proposition with a view of reducing the minimum amount to $20,000.00 per year and upon receipt of your reply we shall bring the matter before our directors.

Very truly yours,

G.P. Jones

General Superintendent

GPJ-B

Keeping the flume running was a problem. The ice built up in the winter and at times the flume froze solidly. Various methods were used to keep the water running, steam heating the water with coal was not cost effective, and the amount of salt needed to keep the water flowing was prohibitive.   Pat Wright photo collection

River, but the Shatford fellow found out they were going to build a dam, so he went down and staked the property, and then when they came to build it he wanted an immense price. In those days, ten thousand dollars was a terrific amount of money but they wouldn't pay it, so they moved downstream.

The flume from the dam to the turbine was 9 feet by 7 feet. I hauled lumber to it by team and wagon. I got the lumber from the carpenter's shop. All this was made in the carpenter's shop, and was tongue and grooved fir. The actual lumber was sawn out here at this side of Sterling Creek Bridge. There was an outfit that had a sawmill in there. This outfit came in from Greenwood and set up a mill out here, near Hedley. Then when the thing was built and completed, they left. There's an old chute just up on that mountain from the dam, and they used to slide logs down. It was a log chute, two or three logs on the bottom and one along the side to keep it in there, and then they'd give it a swat, and down it'd come. At the bottom they'd haul it out to the sawmill. I watched some of those logs come down.

There was one boy who fell in that flume, the McEwan boy. He missed the intake pipe and went through the overflow. Because of the terrific velocity of the water, they couldn't get him in time. There was another error. They found that when the river was in high water they couldn't get power. The high water would not let water escape from the turbine, it just lay dead in there. We had a lot of trouble with that.

This is all that remains of the power dam on the Similkameen River. Dave Taylor, former editor of the Similkameen Star in Princeton, recalled that power from this system was diverted into the West Kootenay grid because of a massive power line failure, and supplied electrical power to the valley, the Hedley complex by day and other customers by night. He recalled running the presses at night because there wasn't any electrical power by day.   Doug Cox photo

There was another reason we couldn't operate it in the wintertime. The ice would freeze in the flume, and you thought you had lots of water - it was really the ice pushing the water up - but there just wasn't enough to turn the waterwheel. At times they had 50 men out there breaking the ice. Then eventually somebody came up with the idea of salt. It didn't do much good. So then they put a boiler down here at the end of the dam, quite a large boiler and they steamed it, but then you couldn't keep up with it. They burned 300 - 400 tons of coal a day there. So you couldn't keep that up."

Early in 1935, when the mine was operated by Kelowna Exploration Co. Ltd., a connection was made with the power supply of the West Kootenay Power Company. The hydro-electric installation on the Similkameen River was shut down. The flume system up Twenty Mile Creek was also replaced by a pipeline, which provided an ample supply of water to the concentration mill at Hedley and Hedley townsite.

Wallace Liddicoat is presently retired in Keremeos after a career in the aero-space industry. He refers to his father, William Liddicoat, as a " Cousin Jack, a person originating in the mining district of Cornwall, England, with a penchant for Cornish pasties, wrestling, singing, cock-fighting, and hand drilling. They were also very adept miners. My father came to work in Hedley in 1908, and eventually became a hard rock miner at the Nickel Plate Mine.

When war broke out in 1914 my father, William Liddicoat, downed tools, as did many miners, picked up his personal effects, and took the train to Nelson where he joined the army. Within six months he was in France as a soldier. In 1916 he was wounded, his legs mutilated, at the Battle of Somme. He convalesced in England for almost a year, where he met my mother Frances May Richards, a volunteer worker in the hospital. They were married in England in 1917. My father returned to Canada on a hospital ship, and my mother came later on a ship as a guest of the Canadian government.

My mother, Frances May, arrived in Hedley complete with a violin, music stand, and portfolio of Brahms, Beethoven, and Bach. She was a musician. Father was convalescing from his wounds, and he had also later contracted the influenza that was world wide after the war. My brother, Art, was born in Hedley in 1918. I was born in Keremeos in 1919, where my father had acquired an orchard as part of a soldiers' settlement program. However, my father supplemented his new orchard by returning to work at the Nickel Plate Mine. A daughter, now Frances May Peck, named after her mother, wouldn't come along till later."

Wallace described his family's stay at the Nickel Plate townsite. " My father worked at the machine shop sharpening steel. We lived in a tent between the railway tracks and Bogus Town. It was quite an experience, but we managed to get by. In 1926 and 1927 Art and I attended school on the mountain. Our teacher was a fellow by the name of Dan Cupid, and his main claim to fame was that his sister was married to Jimmy McLaren, the world welterweight boxing champion at that time. We had to walk from our campsite past the bunkhouses to get to school. There were two bunkhouses, one for the English speaking bachelors and one for the people from Eastern Europe. There were some electric lights along the trail, with bulbs hanging from the fir trees. The winter we spent at the townsite the mine did not close as it usually did when the river froze. We didn't go down the skip on the tram line except to return to Keremeos. At that time there were not many families living at the mine site for any length of time. There were lots of facilities for the single men but not many married quarters."

Bill Liddicoat did continue to work at the mine and did commute via the route to the mine up from the Green Mountain Road. Wallace remembers his father describing how they would reverse the old Model T's up sections of the hill as the gas tank on the cars were gravity fed, and the gas would not flow to the autos' carburetors. Many miners did live with their families in Keremeos and commuted home on the weekends when they had time off. During the week they stayed in the bunkhouse and ate at the cookhouse.

"One crew, Charlie and Alec Saunders, two Englishmen, ran the electric locomotive at the mine. The electric powered motors had one wire, a pole with a wheel, and the railroad track was the ground. Art and I used to ride the loci into the tunnel, to where the ore train was loaded, inside the bowels of that mountain. They had an ore bin blasted out of the rock with chutes to unload the ore. The train backed in till the first car was in line with the chute, Alec would open the chute, fill the ore car, and close it. Charlie, the motor man, would move the train ahead a bit, till they had the whole train loaded. It was certainly not a dry tunnel, and Art and I had to keep our heads down and lay down flat on the motor of the loci to avoid

This was the Nickel Plate school in 1923, with Miss Baxter of Keremeos as the teacher. The Nickel Plate school started in 1905 under Miss. K. Johnson. Bill Orser is the lad with his head through the railing.

Bill Orser photo

touching the wire which was at the top of the tunnel. The driver was safe because he sat in a compartment and had lots of head clearance. We had to lie down or else we might be electrocuted! They didn't have many regulations at that time.

The "powder monkey" used to have his cabin to one side of the mine entrance, and that is where they stored the powder or dynamite. This was where the powder monkey made up all the shots, he cut the fuses to length, and coiled them with the caps. The dynamite was placed in burlap sacks and loaded on a small rail cart and pushed into the mine. He had right of passage when he took this cart into the mine. Art and I used to watch him load the charges and get them ready. The only thing that happened is that we would get headaches from the fumes given off by the dynamite, which was not uncommon. One can rest assured that this would never have been allowed to happen today."

Nickel Plate had its own resort in Strayhorse, or as it is called now, Clearwater or Nickel Plate Lake. The lake had been popular ever since the first miners moved to the area in the early 1900's. The lake was kept stocked with trout, which apparently were a large size. The miners and their families had boats built, which were left at the lake to be used for fishing or a Sunday boat ride. Fishing at the time was recreation and also a source of fresh meat, which was not always possible to have during the days of the ice boxes.

Bill Orser spent most of his childhood at Nickel Plate townsite. As a youth he started working at the mine.

"Looking at the photo of Nickel Plate School, I'm right underneath the teacher with my head stuck out under the rail. I was born in 1918, and I started going to school in 1923. I had to go to school at the age of five because there weren't enough children to get a teacher from

Nickel Plate school children in 1923, with Miss Baxter (later Mrs. Harold Wainwright). In 1906/07 Nickel Plate was one of nine schools in the Similkameen District with a total of 191 students enrolled and a total expenditure of $4204.90 for all nine schools. By 1924, there were 20 schools in the district and 1,549 students, with an expenditure of $42,440.85 for all schools.

Information courtesy of the Ministry of Education

the province. The school was situated on what we call the "high line" going up to the top glory hole. There was only one house beyond the school on that little road going up there. The old road was the old railway track for the trains I grew up by and worked on later, that used to run up and down that track. The school building was originally built as a log house, then it served as a bunkhouse, and it served as a general office for the mine at one time. As they opened up the lower portals, they built a new place down there, so the old place wound up as a school, and later it wound up as a home for several different miners that were there. Mr. Don Smith lived in there with his family until they closed down the mine for the season."

The cookhouse, camp facilities, and some of the general labour work, such as cutting firewood, was the domain of the Chinese men. Bill Orser remembered, "We never did really know how many Chinamen were up above the cookhouse itself. It seemed every time you'd see a Chinaman up there it was a different one! But we did have three or four main stays. Quan was the head one, and he knew me from the time I was about four years old. Whenever I'd show up he'd always say, "Hello Billy. Hello Billy. Come have a piece of pie". The place was just littered with cakes and pies of all different kinds. There was one big Chinaman, he was the one that set up tables. He'd never speak any English whatsoever. He was a huge man, but very old. There was another one who was the dishwasher, general roustabout cook, his name was Yick. There were several others. Every time you'd go down to the kitchen to see Mi, a Chinese friend of mine, there would be two or three washing dishes that you'd never seen before. You would never see them come and you'd never see them go, but we got used to that.

The Orser family had their home in Sicamous where they spent the winter. Most homes at the early Nickel Plate townsite were temporary "tent houses" and the residents moved out when the river froze, which shut down the electrical turbine. Note the milk supply on the running board of the car, a goat!

Bill Orser photo

Then there was Little Joe. He was the general roustabout, he would help at the cookhouse, and he'd go around to the bunkhouses and fill up the different stoves with coal, and sweep out the bunkhouses. He'd always be talking to himself, we always said he was not well, but he did a good job. He'd always carry two buckets of coal with kind of a neck yoke on, like Chinamen used to carry things. He ended up having a big fight with one of the Chinamen and drew a knife. It was Joe Bromley that stepped into that, and Little Joe was taken away I guess, because he had really flipped his wig. We still don't know what the fight was about, possibly Joe Bromley would be able to tell you. Outside of that the Chinese were always wonderful cooks, you could always tell practically every day what it was going to be because it would be the same thing the same day each week. You always knew Fridays because it was fish day. I could never say anything against the Chinese cooking because I really did well on it myself, and I always had a great friend in Quan, Yick and Mi.

The Chinese fellow that was killed going down on the tram car, he was just up there, we didn't know who he was or where he came from, he was one of those that just come and go. I never did find out what name he had. There didn't seem to be any tears shed from any of the others I knew about his demise. He was using a little trolley car, just a little car with boards and a small axle with wheels to run it along the track. You could push them. He was taking lunches down to a track gang. As he was going down the hill, we don't know what happened, I guess he got on the thing and it got going too fast and jumped the track, and he went down through the rocks and got killed.

Eight Chinamen came up to the mine in 1936. They got shovels and picks from the mine

Three generations of Orsers - Samuel, William and Samuel Steven
Orser in 1923.   Orser photo

and went over to the old graveyard on Nickel Plate Road. Bob Alexander, the driver of Kelowna Explorations' one ton truck, was called to go out and pick up all these boxes. The boxes were all filled with Chinese bones; bones of Chinese that had passed on. They were brought over and loaded on our ore train. We got stopped, and they loaded all these boxes of bones on top of the muck on the cars! We took them down, and they were loaded onto a freight car and lowered down ahead of the skip going down into Hedley. As far as I know these bones were going to China to be buried at home in China."

I had the good fortune to interview Joe Bromley at his home in Maple Ridge. He started working at the Nickel Plate Mine in 1924, but his association with the mine was long before that. Joe recalled; "M.K. Rodgers was going to open up and do some work on the Nickel Plate. They hired my uncle, Joe Bromley, as superintendent. He was one of the first superintendents at the Nickel Plate. They got him to gather up a crew of men from Olalla. He did the first work. They drove in two short tunnels in near the Nickel Plate Glory Hole, then they came down and they ran the first main ore tunnel No. 3 underneath the ore body. He was up

A lad's first day at the mine is a momentous occasion. Bill Orser,
then 16, and his father Samuel posed on Bill's first day of work.
Orser photo

there in charge of it for about two to three years. He had a bad heart condition; he couldn't stand the high altitude so he had to leave there.

They went up to that mine by starting from Olalla and Penticton. They had a trail at Hedley, too. They must have brought the supplies up from Hedley by trail. When the Nickel Plate first started there was no road up to Hedley, nothing between Keremeos and Hedley.

My brother, Elwood, was working up at Nickel Plate Mine then. So I put my name in at the mine. He had worked at the Nickel Plate the winter before and was still working there. They had sunk the shaft down 300 ft. and he had worked on that. I got a job through my brother. It was in 1924, April 15th, when I moved up there. I lived down below the cookhouse, in a house. I had to shovel out the trail. There was four feet of snow all the way down, till the 15th of May."

Joe started working at the Nickel Plate Mine in 1924. "The company owned all the houses. There was a row of five houses which they called Bogustown. One had burned down, the

Joe Bromley as a young man moved to the Nickel Plate in April 15, 1924. I had the good fortune to interview Joe at his Maple Ridge home in 1985. He was a very articulate man, who was able to share a wealth of information concerning life and lifestyles at the mine.

Bromley photo

6th one. When I first went to the mine I lived down below the cookhouse, there was one small house. My brother lived up above at the first main tunnel. There had been a lot of buildings there. When they went further down the mountain and drove a new tunnel, they had a motor shed which somebody fixed up into a house. When he left, I moved up into that house. Then later I moved down to Bogustown.

They had a small power plant on Twenty Mile Creek. There was a flume of about three miles long from Twenty Mile Creek down to the mill at Hedley. They had a penstock leading to a kind of turbine engine, and they called it the Goldie engine. A lot of water power went through that and generated the power. That was all the power they had from the time the mill started in 1904 until they built the dam across the river in 1914. They also had a coal-fired boiler and generated some power with steam power.

The compressed air was pumped from Hedley up to Nickel Plate Mine. They had a com-

The date of this photo was July 1, 1926, at the bottom of the old Nickel Plate road off the Green Mountain Road. The Bromleys had left Hedley to go to the Nickel Plate townsite to start a 4:00 p.m. shift. When they reached the present location of the Apex Alpine ski lodge there was four feet of snow on the road. At 9:00 Johnny Brewer from the Nickel Plate Mine came looking for them and helped them through to the Nickel Plate townsite.   Bromley photo

pressor right down at Hedley. When they put the power plant on the river I think they had a bigger compressor. They pumped the air right up from Hedley up to the mine. They first started with a steam plant on the top. The first rock drills that they used in the mine were run with steam. They were called the Burlie machines, and they were made in Sweden. One machine could weigh 150 pounds. They were heavy to drag through the mine up through the stopes and up over ore piles. When I first went to work at Nickel Plate there were still a few of those old machines lying around. A fellow had a hard time to lift them!

They had a boiler just below the mine, and that boiler was fired with wood. They had a bunch of Chinese people cutting dry trees, almost all the way out to Apex over to Clearwater Lake, and all through that country.

They had a steam engine at the lake to pump the water out of the lake when they were running out of Twenty Mile Creek. The creek used to get awfully low in the summer months. They went to work and put a steam boiler and a steam pump on Nickel Plate Lake which pumped the water out of the lake before the dam was built. It was pumped out so that it would run down the creek, from the lake out, raising Twenty Mile Creek to get more power. When I first went there the steam pump was still there, the boiler was gone, but some of the pipe was still there. Later on they used the lake for a reservoir.

I started at Nickel Plate as what they called a mucker, shovelling ore into the cars, taking out waste rocks. There were pockets over the top of the shaft. The incline was a 30 degree incline and at each level were two pockets that had two 2-ton skips in them. We dumped the ore into the pockets and loaded the ore into the skip and hoisted it up. That's what I was doing. Later on I could run a rock drill; then I was a miner. When you got so you could operate a drill, then you were called a miner.

When I first went to the mine, the wages were $1.50 a day. Shortly afterwards they raised it to $2.00 for muckers, miners got fifty cents more, and shift bosses got a dollar a day more.

There were quite a few Cousin Jacks at the mine. They were miners from Cornwall, England. They had a lot of tin mines in Cornwall, and these fellows mined in the tin mines. When they came to Canada they had no trouble getting jobs, they were experienced miners. In the mines all over the country there were known as Cousin Jacks. There were about ten at Nickel Plate Mine when I first started working there.

I worked there until September 1928. The Horn Silver between Osoyoos and Keremeos had been closed down, but did start up again. They put my brother in as foreman. He talked me into quitting at the Nickel Plate and going down there. I worked till the spring, and that was when the stock market crashed. I guess it was the fall of 1929. Silver went down to fifteen cents an ounce, so the company closed down in the spring. Horn Silver, they call it the Dankoe Mine now, had started up in 1914, and has worked off and on since then.

The next summer I went to work at the mill at Hedley; then in November, 1930, the Nickel Plate mine closed down. They thought it was worked out. In the fall of 1932 a new company took over. They put a crew on in development work. I was one of the first to go to work with Kelowna Exploration Co. Ltd. In the summer of 1934 they started up the old mill at Hedley. They started October 15th, 1932 on development work. I became a shift boss at the mine in December, 1934. I worked there until 1943, July 15th.

At Nickel Plate, up to 1931, they had a company store up there but we got our meat from the cookhouse. They used to bring beef up and they had a cooling system. If you wanted meat you went to the cookhouse and got it from the Chinese. They would cut the meat that you wanted. They got the meat from the butcher shop in Hedley, man named Jack Edmonds ran the shop. It would be from the local ranchers, such as Terbaskets, Bertie Allison, Bertie and Jack Edmonds, they used to buy cattle from all over.

I also used to hunt a lot for deer and blue grouse. When I lived on the upper tunnel, there was a ditch of water which ran out of the tunnel and came right through the yard at the back of the house. Alice, our daughter, when she was only about two or three years old, would be playing near the ditch and the blue grouse would come right down for water. They were quite tame. I could shoot them right from the kitchen door. Lots of times we could hunt deer right from the road.

A road was built when I first started. There was the freight road from Penticton, but it was let go for years, it was all culverts and the bridges were all gone, and the road had sloughed in. In the 1930's quite a few people were buying cars down in Hedley. They would leave them at Hedley. They had to go down on the tramway. The Department of Highways started to have the old road cleaned out again some time before 1930. Kelowna Explorations came that way while doing development work in 1931 to 1932 in the summer time. When Kelowna Explorations first started everything was trucked up, when we were on development work. We used a lot of dynamite. Once they packed a whole carload of powder up from Hedley by pack horses. The tramway wasn't running, and the road was snowed in. Charlie Squakin and his sons from the Reserve did the packing.

After the mine got going, they had dances. They had a big dining room and a large room adjoining it. Here, there were long benches to sit on and also card tables and chairs for card playing. It was sort of a recreation room. At meal time, the men would wait in this room until the waitresses opened the dining room doors, then the men would file into the dining room for their meals.

They used to have a little hospital down in Hedley. They used to put on one big dance a year in aid of that hospital. People from Hedley, Keremeos, and Princeton came up the tramway to the dances. In October, the weather was always terrible, so one year they decided to have it in August instead. I think it was the 21st of August. They were hauling people up the tramway all day for the dance. Some even came from Penticton. That night it started to rain, then it turned to snow! They used to dance all night, have their breakfast and then they would start taking them down in the tramway. Overnight it snowed eight inches of wet, sloppy snow. All the women just had light shoes on. They had to wade out in the snow to get on the electric train.

Later on the Kelowna Explorations built a hall up there. That became the recreation centre with pool tables and a bowling alley in the basement, the hall upstairs was for dancing and playing badminton. The Mercer outfit built new houses. There weren't many houses up there. There was a lot of foreigners moved in, mostly Czechoslovakians and Yugoslavians. They built tenthouses.

They used a lot of dynamite, which came in wooden boxes. They would take the empty boxes and put poles up, nail the sides of the boxes for the sides of the house, then put a tent over it. It was awfully cold in the wintertime, with lots of snow - six feet or so some winters. They had the tenthouses insulated and fixed up. They burned wood, coal was brought up by truck. On the tramway, they used to fasten the freight car on the back of the skip; they had a big ring on the back of the skip that hauled the ore down to Hedley. Those skips hauled seven tons of ore. They were eight feet high at the back, and three feet high at the front. There were a couple of awful steep pitches, so they had to have the high back on the skip, otherwise the ore would just roll out as they went down the tramway. People riding up and down had to sit on top of the ore going down. It was quite a thrill to ride down, because they used to go very fast."

Bill Orser recalled some of the events at Nickel Plate. "Mrs. Winkler, nee Mary Brent, really wanted to go to the dance, and Andy, of course, was working that shift. Just before he went into the mine he asked, if his wife wanted to go, would I go down and escort her because it was going to be dark. So I went down and I met her for the first time, and I thought she was the most beautiful lady in the world. Being about 15 at the time, I was real proud to escort her up through the dark from way down by the powder magazine, and take her up there to the dance. I guess she expected that possibly I would take her home, but I think she waited until the shift was over and Andy took her home himself."

I asked Bill about some of the original men like Cahill.

"George Cahill's cabin was like any other cabin that was up and around that part of the area. George was no better housekeeper than any of the other old packers that were there. He used to come out to the mine, and once in a while they'd chase him out of there because he was getting too many free meals in the cookhouse. I don't know how long he lived there. He was a packer, but that trade was dying out too, because roads were getting put in, and trucks could be used. He was one of the first packers to supply the boys that were up at the mine in the late 1800's. He always carried a rifle, and at times he even carried a sidearm, I don't know what for, I don't think he ever used it, but maybe he just thought he'd use it someday. As a prospector, I'm allowed to carry a sidearm in my mining district with written permission from the chief RCMP in Victoria.

I was sent over to register voters at the Hedley Mascot Mine for the voters list. I wanted to get down to Hedley, I had gone through their cookhouse and had as many men signed up as I could. So they said, "OK, you can go down on the tramway." So I had a ride on that aerial tram going down from the Mascot Mine buildings to the concentrator below. I'll tell you, that gives you a thrill if ever there was one! It drops right down over the hill, into the valley."

Mr. Don Smith and Mr. Percy Bailey, both formerly from the Perdue area in Saskatchewan, came to the Nickel Plate townsite via the Bridge Island Mining Co. at Heffley Creek. The men had come from Saskatchewan seeking work during the depression of the 1930's and had found employment in British Columbia. Bruce Hume was in charge of the assessment work for Bridge Island Mining, but told the men he was to be the mining engineer for Kelowna Exploration Co. Ltd., and when he was in the new position would notify the men and give them a job. In due course the men were notified at their homes in Saskatchewan, where they had spent the winter with their wives and families.

Mrs. Mary Smith, Don's wife, recalled that her husband Don and Percy arrived at Nickel Plate around July 1st. There was a heavy snowstorm during this time and the men were swamped with the snow on, in, and around their tent. Don and Percy had driven from the prairies in a small coupe, camping along the way, and had pitched their tent at the minesite until they were hired on. Don worked until fall, then went home to his wife, Mary. Mary didn't come directly to Nickel Plate, Mr. Sam Orser, the shift boss at the mine, found her and

We can thank Arthur Irwin, an engineer who worked at the mine, for some exceptionally clear and artistic photos. Arthur was an outdoorsman and avid cross country skier. He took many artistic as well as informative photos of the mine area. This is Nickel Plate townsite in the winter of 1943.

Arthur Irwin Photo

Nickel Plate townsite and mine was located at the 6000 foot level, a zone usually reserved for ski areas and other winter recreation sites. This is the community centre in 1943, which had a 2 lane bowling alley and pool table on the ground floor, plus a commissary and a hall upstairs for dancing and theatre presentations. It was also a movie theatre when a projectionist and equipment came up from Hedley. *Arthur Irwin photo*

Don a house at Sicamous. There were no houses available at the mine site.

Two years later Don and Mary moved into the original log schoolhouse, which had been renovated and now was a house. They moved in November, however, they did not move directly into their new home, as the new schoolhouse had not been completed. Don and Mary and their family lived for a time in a tent house.

As described by Mrs. Mary Smith, a tent house was not entirely a tent. Most of the tenthouses had log or lumber walls to a certain height, and were then covered with a canvas roof. Mrs. Smith recalled:

"They were not very big, they were light housekeeping rooms in size or smaller, and not too many facilities in them. The company didn't wish permanent buildings built there by individuals, and these were very temporary. Some of them were built with walls the regular height and then just a canvas roof on, they got to be more and more like a little house. Later the company built homes and the people were allotted these houses on a seniority basis."

Don and Mary Smith lived and worked at Nickel Plate until 1944. Mary was able to give us an insight into everyday community life at the townsite at that time.

"When we went there you could use all the electricity you wanted and nobody took any notice, you could heat your house with electric heaters if you wanted. We didn't have electric stoves then, but we used coal. The coal was hauled in from Princeton. Everyone had a coal box close to their house."

Stray Horse Lake, August 10, 1919. The waters of Stray Horse not only supplied the water for the mill and reduction plant in Hedley, it was a recreation facility for the men at Nickel Plate. Fish caught in the lake were kept in a year round ice cave at the mine.   Pat Wright photo collection

Mary went on to describe how they got their groceries and supplies during the winter while at the Nickel Plate.

"There was a wonderful man in Hedley, his name was Tommy Knowles. He was the postmaster. We would get our cheques, and send them down to be cashed, and we would ask him to pay the store, whatever the bill was. (This was the way we did it anyway, but I think he did it for anyone who asked him to). If there was something in the hardware, he would pay these bills for you and send the money back. It would come up the skip. He did that for ever so many people! If you had a savings account it would be in the Post Office Savings. If you asked him to put a certain amount in, he did that, and sent your book up all marked. He was a wonderful man!"

Raising a family in such an isolated remote area may have been a concern for some families, but not the Smiths.

"It was a good place to raise children. They couldn't possibly get into street trouble because there were no streets. There was a very good school. It was very well run. I think the children did just as well there as if they had been in a big city. George Dixon was one of the teachers I remember, and Leafy Dallas was one of the girls who taught there that I can remember.

The Anglican Church had a Ford van which they brought up the mountain for a week or two each year. They had all their Sunday School supplies in that, and all their necessary equipment. The children went to Sunday School each day. They took the children on trips. It was more or less a picnic, but I bet the kids remember more about Sunday School than if they

Nickel Plate Summer Resort, August 10, 1919. Dynamite boxes had many uses at the early Nickel Plate Mine site. They were used for building tent houses, as cupboards and furniture in the tent houses, and as rafts and lakeside furniture at the lake.   Pat Wright photo collection

had been going to ordinary Sunday school. It was a good place to raise children.

I suppose there were 40 or 50 houses in the townsite at that time. When we first came to the mountain any entertainment was in the cookhouse. Any of the single men who lived in the camp lived in the bunkhouses and ate their meals in the cookhouse. They had a piano up there and there was generally someone who could play. Bill Morgan played violin for us and Mrs. Taylor played the piano. You could get up a dance in half an hour, you had free music, and somebody always managed to bring something for a lunch. It was a good life there, and we were all good friends. It was really a self contained place. You didn't need to go away from there."

Mrs. Smith went on to describe how they got their groceries: "We'd send a letter down to the stores, McDonalds had a store and Farmer Bush, and ask for whatever groceries we wanted. They would pack this all up and send it up the skip, and it would be charged to your account, which was paid by Mr. Knowles on payday. The groceries were put in a cardboard box and taken over to the skip, delivered to the office, and you picked it up there.

We didn't travel up and down the skip too often. My husband thought it wasn't a safe place to go with three kids. We did go down a few times. You sat on top of wet ore and hung on when you went down, and it was down too! It stopped at Central and had different cables hooked to it and went on the rest of the way down."

Don and Mary had a family of three girls before they went to the Nickel Plate. Their son was born at the townsite. "There weren't any medical facilities at the camp, but when our son was being born the doctor was sent for, I believe it was a stormy day, the 20th of April and he didn't get there till 3 or 4 o'clock in the afternoon. By that time my husband had delivered

Eileen McCallum in the doorway of a tent house at the Nickel Plate Mine, 1937. It was the largest tent house in the camp. The tent was sold to Mary and Howard Graham of Keremeos when Eileen, husband, Bud, and son, Jack, returned to Copper Mountain along with their dog, Rex. Eileen said they were quite warm - there being a large stove in the middle - it was a cold winter as well.
photo courtesy Evelyn McCallum

our son. My neighbour lady was there, but she was watching out the window for the doctor to come!

The cookhouse was open to the people who lived on the hill. We could go and get meat but no groceries. It was put on the bill and you paid for it when payday came. The cooks were Chinese and they cut their own meat; they could cut you a steak an inch thick or a foot, whatever you liked. Quan was the chief cook when we were there, and even after we left. Quan used to go to Vancouver when he had his time off, and would come back a little richer or a little wiser from playing fantan.

Mr. Alex Gough used to bring moving pictures up there, he would put a show on twice a day so that everyone could see. The people who were going on shift could go in the morning, the ones that came off shift could go in the afternoon. He came very faithfully every week.

The Nickel Plate Mine orchestra in 1939 featured: Bill Morgan on violin; Eileen Taylor, piano; George Dixon, violin; Cliff McDonald, drums; and George Ferguson, banjo. This orchestra played for dances at the community hall and special occasions when Nickel Plate Mine bachelors hosted the "Bachelor Ball"

There was a baby born in a Hudson Terra Plane near the rock bluff on the way to Princeton, the girl lives in Oliver now with a family of her own. She has always been called Terri Kelly even though Terri was not her name. Bill Orser recalled the birth of this girl. "It was the winter time. They couldn't get Joe Kelly's car going in the middle of the night to take Mrs. Kelly to the Princeton hospital - I'd gone half and half with him to buy the car in the first place. So Theo Prest and his wife used their brand new 1935 Terra Plane car and took her down. But they only got as far as "Maternity Curve" where Mary (Mrs. Kelly) gave birth to a baby girl.

Mrs. Theo Prest came out of Vancouver. She had had a very interesting life in Vancouver. She married Theo Prest who was the manager of the ore bin area. One time she had to doctor a man who had a little accident. He had caught his foreskin in the zipper of his pants! After much trouble, they got him over to a bunkhouse, and they all kind of gave up, 'cause every time they tried to touch him he'd just about go through the roof. So they brought Mrs. Prest around, and she said, "Well, she'd seen these things before." She said, "This won't hurt much," and she gave it a zip and he was free! He really just about went through the roof then."

Mrs. Ann Brown and her husband and two sons, Murray and David, moved to the Nickel Plate in 1943. She recalled that their first house was just below the Community Hall, and one had to climb steps to get to that level, then a flight of stairs to get to the track, and then on to a trail to the school. Along the track about a quarter of a mile were the office buildings and

Nickel Plate Ladies Bowling League, 1949. Back Row L to R: Tina Loeppky, Hilda Palumbo, Millie Dannhauer, Berta Forgaard, Anne Brown, Phyllis Hesketh, Vera Fenton, Kay Hesketh. Centre Row: Anna Richards, Mary Graham, Peggy Hendsbee, Rae ---, Gwenn Larson, Elsie Leipert, Mary Emery. Front Row: Nonnie Casey, Betty Koenig, Marion Allen, Lydia Shaak.   Sid Dannhouer photo

the cookhouse and the main portal of the mine. The bunkhouses for the men were in rows on two levels all neatly painted. The view from this level was very beautiful. On a clear day one could see the river far below.

"The cookhouse was an enticing place to visit. Sometimes the head cook, Charlie Moy, would ask my husband and myself to bring the boys and come for a meal. Their father warned them to be sure to eat everything on their plates or Charlie would be very offended.

My husband was the accountant at the mine and served part time as the post master. On the days of doctor's visits, which were once a week, he made the necessary appointments." A Post Office was established at the mountain townsite November 1, 1945, known as Nickel Plate, British Columbia.

"One winter's night I was called to see a sick baby. The parents were newcomers but they knew I was a graduate nurse. The child was very ill, and everyone tried to get a car to start to run me to the doctor in Hedley or to the hospital in Princeton. I gave the wee mite mouth to mouth resuscitation all the way, but it was too late. The "Hill" was a poor place to be ill. My own son, Murray, contacted pneumonia one winter, and had to have shots of penicillin. The doctor made a special trip up to see him, and I spent some anxious nights with him, also.

In the Community Hall we had a large main hall where dances and meetings were held, and a lower hall where there was a commissary and a bowling alley. The ladies formed a team, and we had many a good evenings up there. The men took advantage of the bowling alley when they were off shift. Badminton teams often came up for tournaments with our

This is a Sunday School class at the first log school at the Nickel Plate townsite. Doug Cox photo collection

players. They didn't understand why their serves always went out of the courts till we reminded them to "go softer" as we were at 5800 feet elevation here. A distinct advantage for the house team.

At Christmas time we made curtains for a stage and put on the school concert. On Sundays there was often a church service, any denomination was welcome, and someone was always ready to play the piano. The minister I remember most was Max Warne, who came up on Saturday evening. There were cars on the hill by then, and he played for the dance and preached a sermon on Sunday morning. His violin always came with him. I enjoyed hearing him. We later purchased his 1946 Ford car.

The manager, when we first came to Nickel Plate, was Alex Shaak. Luther Yantis, a later manager, was driving up from Hedley one warm day and saw a dead rattlesnake on the road. He picked it up and put it in the trunk of his car and went on home. As he put his car in the garage he noticed an agitation between the lining and the roof of the car. The snake had come to life! He had quite a time to coax it back into the trunk where I expect it met a quick demise.

On one occasion another nurse and myself were called down to the guest house in Hedley to nurse Mr. Mercer, the general manager of the mine. We subsequently went with him by train to the Mayo Clinic in Rochester, and spent a day or so looking around the various buildings before coming home.

There were many more things that happened up on the "Hill" and I wish I could recall them all. I still retain friendship with any ex-Nickel Plate folk who have settled in this valley and at the coast. It does seem that "old friends are best friends", especially among mining folk."

Harry and Mardy Wood, relatively late comers to the Nickel Plate, give an idea of life at

Main Street, Nickel Plate Mine site in 1943. The large building is the Community Hall which combined a 2-lane bowling alley, a gym, stage and small store. George Pizzi and John Allison hauled the bowling lanes for the Recreation Centre, up the Nickel Plate Road, in the early 1940's. The trucks they used were a 1938 and a 1940 International. They had one bowling lane on each truck. The lane had to be in the center of the truck deck so they would have enough room to make the corners on the road up to the minesite. The lanes had to be hauled on edge so they would not buckle. To hold the lanes in place they built a frame on each truck. There was about 20 feet of bowling lane over the front of the truck and the balance over the back of the deck. Information courtesy Lenard Wheeler.   Arthur Irwin photo

the townsite. He recalled, "I started work at the mine in late August 1949, moving from Greenwood, B.C., where I had been employed at the Brooklyn Stemwinder Mine. My wife, two boys, Daryl and Gar, and myself moved to Nickel Plate and into one of the tent houses on the lower road. Mardy's brother and his wife and family lived up the road a little way, so with their help, we soon got to know a lot of the residents on the hill. The winter of 1949/50 was a hard winter, but we had lots of wood cut, and so didn't suffer any. The snow was piled high, but it didn't dampen the good times at the hall and house parties. Cliff MacDonald got an orchestra together, and so we didn't have to go down the hill for our entertainment. The first big thrill after moving up was a ride on the ore train and then down the tramway to Hedley, very exciting indeed. There was no T.V. and very poor radio reception, so everyone had to get together and help out with entertainment. The men worked hard underground, and then found time to work hard making ice and trying to keep the snow off it, so as to get in a bit of skating. We made bobsleds and had a great time riding them down the road, but as cars got more plentiful, they had to put sand on the road and that spoiled the bobsledding.

Verita, our third child was born in 1953. We had moved to a bit better house by the tennis court by then, so we had a bit more room. I might add here that our fourth child and last,

Lynnette, was born in 1956 in Penticton, after we left the Nickel Plate.

By the fall of 1954, there were rumors that the mine may have to shut down for lack of ore. We didn't want to believe it, but by the spring of 1955 it was announced by management that the mine would close by fall. It was quite a blow, not only because of being put out of work, but because it would break up a very good community.

We will always remember Nickel Plate as the best place we have ever lived. Mainly because the people were all so friendly. It was with great regret that we packed up our things, and in July 1955 we moved to Penticton, where I had found work with Parker's Industrial. Since that time we have helped to organize the first Nickel Plate reunion which we enjoyed very much, and have since attended three more. My family and I are presently living in Hope (since 1961) where I am employed by Anderson Construction Company.

This is a poem written by Mardy Woods' mother, Mrs. Beatrice Allen, about Nickel Plate. She thought it was such a beautiful place. She used to come and visit often.

Away upon a mountain
Above Similkameen
Was the nicest little mining camp
That we have ever seen.

The houses they were varied
Stairways joined street to street
And the people living up there
Were grand to meet.

The road was noted far and wide
For steep and twisting turns
T'was not the sort of joy ride though
For which a person yearns

As many guests who climbed the hill
For courage they did look
They stayed up there for days and days
Afraid of coming back.

The men worked hard down under
But didn't seem to care
For richer than the gold they mined
Were friendships made up there.

The mine closed down for lack of gold
And neighbors moved away
The camp is now a ghost town
We think that's how t'will stay.

A last verse was added for the reunion held in Hedley in 1961:

And so we've come to Hedley
To help to celebrate
And meet again the friends we made
Way up at Nickel Plate.

These were some of the houses at the Nickel Plate minesite built during the time of Kelowna Exploration Co. Ltd. (1934-1955). When the mine closed in 1955, the houses were sold and moved to other communities.   W.V. Ring photo

The Nickel Plate townsite, although very close to Hedley, especially when the new road opened, had a lot of recreation facilities. The community hall had a two lane bowling alley and a billiard room. The complex also contained a small store and the dance hall. In summer there was a tennis court that boasted a wooden floor, and in the winter the citizens usually flooded an area that was a skating rink. As in any mountainous area, the children found a suitable slope and went sleigh riding.

Randy Manuel is presently the director of the Penticton Museum. His recollections of the Nickel Plate townsite are through the eyes of a six year old:

"My reason for being up at Nickel Plate was that my mother had a sister and a brother-in-law living at the townsite. Mickey, nee Beaton, and Stuart Munro had been living at Hedley since 1938, and in 1950 Stuart was transferred to the office on "top", where he was office manager, timekeeper, and first aid man. Mickey and Stuart had two kids who lived up there and had to go to school. My mother did enroll me at the Nickel Plate School, which was impossible in Penticton as I was too young, according to the Penticton School District. Once I was enrolled I could transfer back to Penticton as a regular student. That's how I started out, going to a little two room school up on top. I believe the teacher lived in the basement of the school. From a kid's point of view it was a neat place, because it had a sand box and other neat toys.

On Saturdays we kids would go down to the commissary, which was below the school and farther to the west. The commissary was located in the Community Centre, which had a two lane bowling alley at the ground level, plus a small confectionary and the Post Office.

This 7 foot 9 inch cougar was shot behind Nickel Plate in 1948 by Howard Graham and Sid Dannhauer. Cougars were a problem in the winter, bears in the summer.   Sid Dannhauer photo

Upstairs was a badminton court with a stage at one end where plays could be put on, and also where the 16 m.m. movie played twice a month. The Cubs and Brownies met there also. To get to the commissary one went up and down flights of stairs, which were roofed over because in the winter it made them easier to keep clear of snow. Below the school, but above the commissary, was the tram line.

On some Saturdays we would often take a picnic of wieners and marshmallows and hop a ride on the loci. I imagine Workman's Compensation would go wild if that were to happen today, but as kids that's what we did! We would ride down to the skip, and while the tram cars were being unloaded we started a bon fire and roasted our wieners and marshmallows. Why we didn't burn the hill down I don't know! When the train was ready we would ride back on the loci, or walk back poking into abandoned cabins and mine adits along the way.

West of the school near an abandoned ore dump was the ski hill. The ski facility consisted of a motor which operated the rope tow. All one had to do was strap on a set of boards and ski up and down. The tennis courts were flooded in winter to serve as a skating rink.

During the summertime my mother and I would go over to the Nickel Plate frequently, and I would spend some time with my cousins. To get to the townsite we went up the road from Hedley. It was great in the summer. Other times were different. I remember my dad had a 4WD Willy's Jeep, and that road gave one the willies also! The road wasn't bad until you reached the rock cut near the top. At this point the road was about 10 feet wide and held up by log cribbing. After that was a fifteen hundred foot drop to the talus slope below. On this particular trip the jeep started to slip back on the snow, caused by the altitude and temperature change. I can remember my mother being rather excited as we started to slip backwards. You could barely see the light of the cars on the highway, several thousand feet below. My dad chained up all four wheels and we made it to the minesite. Why more people didn't fall off that road, including my uncle, who used to "exercise his right arm" too much, I don't know.

When the mine was being shut down the furniture from my uncle's house was bundled up and put in an ore car and taken down. Included was a grand piano, which sits in my front room forty years later. It has scars from being somewhat battered, from being lowered into an ore bucket.

The mine was a great place for kids to play, because there were so many things going on. One thing that we as youngsters did was haunt the cookhouse, which was on the mainline track near the office and the bunkhouses. Quan and Kim were the cooks. Quan was taller than Kim and both spoke very broken pidgin English. My dad seemed to understand them very nicely as they always had a mason jar of whiskey handy. As kids we were allowed to go in and have cake or pie or anything we wanted to eat or drink. Often we walked back to the house along the tracks with a hot apple pie sent home to Stuart and Micky for their evening supper.

If Dad or Stuart were to go down after supper for a game of cards, out came the mason jars. My father told a story of Kim reaching up for a jar that was to be used as a drinking glass. The whiskey was poured from one large jar into smaller jars. The jar that Kim brought down had a spider web and a dead spider in the bottom. Kim put his finger in the jar, wiped out the spider web, blew out the dead spider, and poured in some fresh whiskey.

My father, Al Manuel, worked for Interior Contracting on Fairview Road, at that time owned by the Hatfield family. They contracted to the railroad to do maintenance work, which is why he is sometimes associated with the Kettle Valley Railroad. Interior Contracting did the repairs to the mill when it burned in the late 1940's. Occasionally Interior contracting was brought in to improve the Nickel Plate Road.

By 1955 the Nickel Plate Mine had folded, and my uncle, Stuart Munro was the last company officer to leave the mine site. The houses were taken down the mountain and reused, either in whole or in part. When we, as teenagers, went up to the site in the 1960's, all there was were the foundations and outbuildings of a once thriving townsite."

10. GOMER P. JONES

"Gomer Phillip Jones, prominent figure in mining development in B.C. since 1900, died in Vancouver, Thursday, March 25, 1951." This is taken from his obituary as it appeared in the Vancouver Sun.

"As general manager of Hedley Gold Mining Co. Ltd., which operated the Nickel Plate mine at Hedley for 30 years, Mr. Jones was particularly well known. He was also past president of the B.C. Chamber of Mines, past chairman of the Vancouver branch of the Canadian Institute of Mining and Metallurgy, and a charter member of the Association of Professional Engineers of B.C.

Born in Australia, Mr. Jones started mining there at the age of 14, and by attending night school, earned his chemical engineering degree before he was 21. He later received a mining engineering degree from the Bendigo School of Mines. In 1892 he went to the United States, and in 1900 arrived at the Nickel Plate mine as mine superintendent. He became general manager of the company nine years later and held that position until operation was suspended in 1930. He retired in 1932.

Prominent in Masonic circles in B.C., Mr. Jones was past worshipful master of Hedley Lodge, AF and AM, and a past patron of No. 2 Chapter, Order of the Eastern Star. He leaves his wife, one son, Gomer P. Jr., at Trail, and a daughter, Mrs. C.E. Prior, Los Angeles."

G.P. Jones and wife leaving Hedley in the 1930's. G.P., an Australian, was instrumental in the development of the mine and reduction plant. Gomer P. Jones, who was to be connected with the Nickel Plate for so many years, was engaged by Rodgers as mine superintendent, and he arrived at the camp in August, 1900. Mrs. Jones and their daughter Avonia came a month or so later, and took up residence at the Nickel Plate. Pat Wright Photo collection

G.P. Jones and his wife at Stanley Park in Vancouver in the 1940's.
Herbert Jones Photo

Ioco, B.C.
Jan. 24, 1986.

Mr. Doug Cox,

Yes, I am the nephew of the late G.P. Jones. I would say he was a worker even when he was promoted from Mine Superintendent to General Super. Downtown in Hedley he continued to be in the thick of things in all departments and activities in the town, his predecessor as General Superintendent sat on the veranda of the office a lot of his time and never worked like my uncle.

I'll call my uncle G.P., as that is what I called him most of the time I had any contact. He came from Australia as a young man in the 1890's to New York, U.S.A. where he started and operated a bicycle shop. He married in New York and had one child there. G.P. read in a New York paper one day an ad for a mining man to operate the Nickel Plate Mine. He applied for the job and got it, and lost no time getting there.

Leaving his wife and daughter in New York much against their will, as there was no accommodation at the mine for men, let alone women, but his wife and child arrived soon after G.P. did, and they lived in a tent house until a log (no lumber) house could be built for them.

G.P. was a lucky and fortunate man. He said everything he touched turned to gold. My father, G.P.'s brother, came to the mine as a shift foreman under ground (1903). He had a lot to do with finding a glory hole - very high grade, (around $4 a ton at least) I've forgotten which it was - they used to sweeten the low grade ore (you see I'm 87 years old and don't remember that far back too well). G.P. had a brother, who came out to the mine also; these two men were of tremendous help to him, who when he died, was a very wealthy man.

The only time G.P. never got his way was with Duncan Woods, who owned the Mascot Fraction in the Heart of the Nickel Plate Mine. G.P. worked hard to get it, as it would have been most beneficial to him, but Duncan would never let him have it. Duncan was a friend of mine and my family, and if I were sawing a log for firewood he'd come along and say, "You are doing that the wrong way". "How would you do it, Dunc?" I would reply. Dunc would take the saw - a 4 ft. crosscut - and saw the whole log up, that was okay by me. Other times Dunc would meet my wife in downtown Hedley, and she'd have our three young girls with her on a sled. Dunc would pull them home for her. Yes, Dunc was a character, but always had the time of day to talk to us. But when it came to G.P. it was nil. I never found out the trouble in over 15 years I knew him, Dunc said he'd see the grass grow over Hedley's streets before "Jones", as he called G.P., got the mine. When Dunc finally sold the Mascot, G.P. had retired and was living out here. It's said that the day the new Mascot Mill got started, Dunc who was ill in Penticton Hospital, died.

I'm enclosing the address of G.P.'s granddaughter, who was a favorite of his, and I am told has some information and pictures that might be useful to you. I'm also enclosing some clippings from a Vancouver newspaper about G.P. and a snap of G.P. and Aunt Minnie, his wife. I saw a news item in last night's Sun about the mine, and it looks like the pit might be the site where the mine site used to be. I think it was once Bogus Town where our family lived. Would like clippings back, no hurry.

Yours truly,
Harold Jones

The following article was taken from the Similkameen Star in Princeton, Wednesday, January 25, 1939.

TWO DIE IN ROCK SLIDE

Graphic Story of the Hedley Disaster

Five Houses Struck as Tons of Rock Hurled Upon Sleeping Townsite

Two persons are dead, and property damage estimated at several hundred dollars was occasioned as huge boulders, loosed by expanding ice, crashed 1200 feet down the precipitous west side of Stemwinder Mountain, onto the sleeping Mascot townsite below, about 1:30 Tuesday morning.

THE DEAD:

PETER STRAND, 54 or 56, native of Norway, resident here about 25 years. Employed as watchman of slag pond, Mascot gold mill Hedley, about three years. Unmarried.

JOHANNA SOPHIE GREEN, 54, born 26th April, 1885, Norway. In Canada about 30 years. Survived by husband and two married daughters, in Princeton. Entered service of Strand as housekeeper just before Christmas.

Curly Phillips, Bill Corrigan, and Art Rowe beside a rock from the Stemwinder which killed two people as it went through a house. The Vancouver Sun's caption to this photo read: Tons of rock, in pieces weighing 20 and 25 tons, crashed 1000 feet down the southeast side of Stemwinder Mountain to sleeping Hedley, B.C. early Tuesday morning and brought death to Peter Strand and Mrs. Johanna Green. The picture indicates the force of a 20-ton boulder that nearly decapitated Strand and Mrs. Green's neck as they lay asleep in the house. The boulder tore through the home and came to rest a distance outside. The rock, which killed the two, is shown in the foreground.

Dave Innis photo collection

Pete Berkman, the Great Northern Section foreman, and his wife standing by a boulder which stopped blocking in their back door.   Dave Innis photo collection

Five buildings were struck by huge boulders, loosed from a gaping socket high on the summit of the west slope of Stemwinder Mountain, which rises almost perpendicularly above this section of the town, and catapulted in gigantic leaps, at terrific speed in a 1200 foot drop. Twelve boulders came down, averaging twenty to twenty-five tons, and showers of smaller rocks accompanied them.

Wrecked buildings include:

STRAND COTTAGE: Huge rock smashed through bedroom wing, killing two sleeping occupants.

BERKMAN PLACE: Two big rocks ploughed through yard, demolishing woodshed. One came to rest a sparse three feet from back door. The other cleared house and glanced off pine tree.

MOORE RESIDENCE: Rock smashed bedroom just after Betty, 14, and Helen, 12, had evacuated. Damage considerable.

RUDOLPH HOME: Sitting room demolished by huge rock while family slept in bedroom. Damage considerable.

TURNER RESIDENCE: Bedroom plundered by boulder. Family were in front room. Damage considerable.

In addition, damage was done to the McCourt garage and an 800 pound stone struck the Tremaine garage.

Fear of further slides caused the entire section to be evacuated tonight, as residents were jumpy and unnerved. It was, however, the first serious slide in 40 years, although minor slides are common.

The Jack Moore residence in Hedley after the rock slide from Stemwinder Mountain January 24, 1919. The Vancouver Sun January 25, stated: ...photo shows what remains of the bedroom of two of the four children of Mr. and Mrs. Jack Moore. The children were asleep in the room when the slide started and ran from the room just before a huge boulder crashed into it.

Dave Innis photo collection

Dr. Daniel McCaffrey, Princeton coroner, announced after visiting Hedley Tuesday afternoon that an inquest will be held Thursday at 2:30 at Hedley. An autopsy of Mr. Strand and Mrs. Green will be held Wednesday morning at the Princeton morgue, where both bodies lie.

Strand was killed instantly as he lay asleep, Mrs. Green was unconscious and died on the steps of Dr. Wride's office where she was being carried for treatment.

The rock which dealt double death ripped with razor sharpness, borne of terrific speed, a clean hole through the frame building and came to rest twenty-five feet from the house. At the southerly end of an area of about one acre, it was one of the first to come down during the terrific 15-minute bombardment.

Startled by the roar which was heard for miles around, a crowd immediately gathered to witness the devastation. The clang of the fire bell accentuated the dramatic atmosphere. Fire was a distinct menace, though only one minor outburst threatened, when an electric stove in the Rudolph place collapsed and short-circuited.

Victor J. Creeden, secretary of the Hedley Mascot Mining Co. was en route from Vancouver tonight to investigate the company's losses. Most of the damaged houses are owned by the company. It is understood that insurance of this kind was not carried. Residents too, lost heavily in damage to furniture and private property.

It is believed that alternate soft and freezing weather were responsible for the disaster. Water is presumed to have gathered in the cracks, and its expansion while freezing caused the fissure.

Flooding devastated Hedley in 1948 as it did many communities. The house at the left is a good example. Levees and dams were constructed to contain the waters of Twenty Mile Creek which were swollen with flood waters caused by a heavy snowfall and late spring. Ed Aldredge photo

Stemwinder or Striped mountain (so called from its variegated contours) rises abruptly from the Twenty Mile valley on the northwest, and slopes gently from the Similkameen Valley on the north. It was prominent in the news as the location of the notorious Hedley Amalgamated mine.

The coincidence of a "Major Bowes" entertainment in the town hall was perhaps a happy factor, in that many had attended and were either still up or had gone to bed late.

There is diversity of opinion as to the exact time of the avalanche. Neill McLeod, who says he stood on his verandah as the rock flashed by not ten feet away after killing his two neighbours, says that the slide occupied a fifteen minute period between 1:15 and 1:30 a.m. Mrs. J. Ferko, who had not retired, agrees. Other neighbours are vague. Const. F. Lines says its was almost 2 o'clock when he arrived, one of the first on the scene.

McLeod says there was a roar like thunder, and that the huge rocks shot through the air in mighty leaps like a bullet.

It was little short of a miracle how some of the residents escaped. A huge rock stopped dead a scant three feet from the Berkman door.

Betty (14) and Helen (12) Moore, daughters of Mr. and Mrs. Jack Moore, were asleep in a downstairs bedroom when awakened by the rumble of rushing rock. As they went upstairs to rouse their 11-year-old brother, a rock splintered their room.

The Rudolphs were asleep in the bedroom when a rock shattered the living room. Mrs. Rudolph suffered three cracked ribs scrambling among the wreckage for her pet cat.

Mrs. Floyd Turner was awakened by the noise and roused her husband. A few minutes later their bed was a crumpled mass, pushed into the front room.

Mrs. McCall awoke and roused her twin sisters, her guests.

Mr. and Mrs. J. Alcock had moved from the neighbourhood a few hours before.

The terrific speed attained by the massive boulders in their 1200 foot plunge can only be conjectured. Evidence is graphically seen in the cleanness of the paths they cut. The huge boulders did not crush; they sliced and splintered.

One huge fragment hurtled over fences, in 150 foot hops, churned through the vacant yard between the Berkman and Heeney places, struck the street and bounded almost 200 feet to come to rest harmlessly in the alley.

A naked cup-shaped gouge high on the marbled face of the hill clearly tells the story. The rock is of hard, igneous formation.

An eager, night-capped throng quickly gathered to watch officials and recruits recover the bodies and survey the devastation, under supervision of Cost. F. Lines, resident officer. B.C. Police and Dr. G. E. Wride, both of whom arrived almost immediately.

During the day crowds visited the forlorn scene, visitors coming from as far afield as Penticton despite falling snow and difficult roads. Hedley has about a foot of snow, considerably more than Princeton.

Old timers recall a slide of some seriousness near the same place about 30 years ago, but say no great damage was done. Many of the houses were comparatively new.

The Strand bedroom was entirely wrecked, though the walls are intact, save for a gaping hole. The bed was smashed to fragments.

Everybody in Hedley speaks very highly of Strand, whom they call a prince of fellows. Mrs. Green, an old friend, joined him just before Christmas. She was preparing to leave in a day or two to attend a wedding party at Copper Mtn.

She is survived by her husband, Charles A. Green, and two daughters, Mrs. A. Clarke and Mrs. J. Faulson all of Princeton. Funeral arrangements have not been completed, but interment will probably take place Friday at Princeton.

The Similkameen Star considered the January 1939 rock slide at Hedley as one of the Similkameen's major disasters to that time.

SIMILKAMEEN'S FOUR MAJOR DISASTERS

- Copper Mountain Holocaust - 17th March, 1928. Nine men killed, 15 injured in bunkhouse fire.

- Blakeburn Explosion - 13th August, 1930, 46 men killed.

- Copper Mtn. Shaft Disaster - 4th August, 1937. Seventeen killed. One died later. Cage crashed down 600 foot shaft.

- Man was killed when boulder crashed into his cabin in Granite Creek canyon near Coalmont in winter of 1936.

12. MINING

Bill Orser recalls some of his memories of working at the Nickel Plate mine.

"I spent quite a bit of time in the hoist room, as a kind of wiper, wiping up the greases and the oils, and keeping the dust down, keeping the place tidy, and washing the cement floor. I used to get a big kick out of watching the dials as the drums were letting out the cable of the skip going down 1500 feet on an incline. I'd watch the hands on the dials go, and the operator could stop the skid right on a button if he wanted to. Miller Kirkpatrick saved my father's life with knowing exactly where his skip was down that track. Dad had loaded some starter steel, 3 foot long drilling steel, on the bail of the skip - he was the shift boss at the time - and foolishly got up on that steel and was going to ride up to the top. The track going down this incline, which was called the "Dickson Incline", was quite rough, so the steel started shimmying on there and slid out from underneath him, and he fell off that skip. He hit in the middle of the track, and he hung on to the cable, jumping across the different places that the cable used to ride on, which was hurting him real bad, and with a superhuman effort, he threw himself to the side, letting go of the cable. Unfortunately, he hit a timber and he had

This is an underground cavern being mined out deep in the bowels of Nickel Plate Mountain during the 1920's. The driller, top left, is using a type of mining drill used very early in the mine, which started in 1900, and is now being mined by open pit methods. The men (centre) are wearing ties, the men with the wheelbarrow are muckers. The photo flash was produced by burning gunpowder, which accounts for the flash to the right.   Doug Cox photo collection

Carl Loomer suggested the drill in this photo is a Sullivan machine. He said, "It is clamped to the bar and that has a thread on each end. You screw it up and make it tight. Men used to pack that thing around the mine. A fellow by the name of Joe Bromley used to put the whole thing on his shoulder and go. Also in the photo is an oil can, oil that they drained out; when it got air all through it got all emulsified quite quickly; it did not last very long. The machines would go through a 45-drum of oil in probably two weeks. Mind you, they filtered this back and reused it. But when it got so frothy that the mine was afraid it was no longer lubricating they would discard it." The year scratched in the wall is 1936, which gives an indication of the year the photo was taken. The lad in the white shoes is possibly Jack Stocks, who often helped his father on photographic assignments. The square tin contains oil which lubricated the drill. Hardhats were not mandatory, and many of the older miners did not wear them. The hat, however, held the carbide or battery lights for the miners.

Stocks photo

his stomach up against the timber as the skip slowly came up, and the trunion on the rear wheels of the skip caught him in the back and tore up five of his vertebrae, causing a horrible injury. This happened in 1935. Miller Kirkpatrick, who was the hoistman at the time, said he felt that on his controls and stopped the skip after it had, of course, passed about three feet, which was a little too much for dad's back. But dad was able to stand up, and he got up and rang nine bells on the cord that hung there. He pulled the cord nine times, which was a danger/accident signal. So, of course, it wasn't very long before everybody seemed to get there. They picked him up and brought him up to the top, and he spent the next two-and-a-half years in St. Paul hospital in Vancouver under the care of Dr. Star of the Workman's Compensation Board. He did walk again, and he got a farm going at Oliver. He worked at the

Sam Orser and associates are preparing to blast. The holes are all loaded up with powder, the fuses are hanging out of the holes. Sam Orser is holding the powder punch pole used to jam the dynamite into the holes. In Sam's coat there is a piece of fuse called a "spitter", which is notched with a knife about every inch or so. One can light one end of it and the blaster will "spit" his holes starting with the centre holes, then the knee holes, then the upper holes, and the lifters last. The lifters go last, which throws the muck around and breaks it up. When the blaster has everything lit there is nothing but a pall of blue smoke from the fuses, and he is just gumbooting out of there.   Bill Orser photo

Oliver Sawmills as a painter and blaster, but lots of times he had to take days off to get his back back into shape. He started off with full compensation, at about $200 a month, which was not too bad at the time. But it kept on dwindling down and dwindling down; his last cheque was $36 before he died in 1971.

Operating a mining drill needed technique. You'd have to grab hold of the bar, that was

This underground trestle was at the 400 foot level of the Dickson Incline. Built to cross a stope, the bridge was 108 feet long and 37 feet high at the highest point, curved and covered with ties and a 6-foot deck. It was built by Sid Dannhauer and a partner in less than five days for $275. The photo was taken with photo floods. Sid Dannhauer Photo

the heavy part, and there was the knuckle on that bar, and there was a water hose coming in. There was also an air hose and, of course, the drill steel fitted into the throat of the machine. There were wedges up on the top to wedge the drill in there good and solid. There was also a hole up on the bar that you could take and put a sprague in and turn it, and jack the drill against these wedges. Then there was the knuckle, and your machine mounts onto that. You can swing that thing right completely around. Then you turn the handle, which turns counter-clockwise, to feed the steel in. The machine rotates the steel and hammers at the same time. The water comes up and goes down through the centre of the steel right into the bottom of the hole, bringing all the cuttings out. That's why a man looks like Sandy Brent does in the photo on the next page, with his face dirty. The water would be coming out of the hole.

How far apart the driller drilled the holes would vary. Sometimes they'd be about 14 inches apart. The first two holes will go in and meet. These are the centre holes. Then you'd probably put another one in from the bottom, going up a bit. Then you may put in a few more at different angles. Then you put the dynamite down there all in one spot. You time it, so that all four holes will go at the same time. When the EB, electric blasting caps came in, especially what we called the 'millisecond' cap, they were wonderful. You could now set zero time on the

Alexander (Sandy) Brent, hard rock miner, somewhere in the Nickel Plate Mine. Born and raised in the Okanagan, he worked as a rancher, miner and construction worker, Sandy was a wealth of information on any topic. I often would take Sandy up to the mine site, or anywhere in the area, to learn about the early Okanagan. A tape recorder was usually on the front seat of the truck.

Brent family photo

millisecond, those four holes in the centre would be 'zeros' , the next holes on the outside would be number ones, and the next ones out there would be number twos, and so on. Then the bottom ones would be set at five or six; that's the one that would heave the muck all over to one side of the drift, giving you room to get in there with very little mucking to do, and making it easy to get your steel plate down for the next blast. The millisecond caps allowed the first four holes to go off, pulling the plug out, and the next would just cave into the centre of the hole.

A 'bootleg' is a hole that has a charge that hasn't gone off. That usually happened on what

Sid Dannhauer at one of the glory holes, which were a result of
early open pit type mining.   Sid Dannhauer photo

the drillers called 'lifters'. There were usually three lifters drilled in the bottom of the drift.
What would happen is that number two goes off prematurely, ahead of number one, and cuts
the number one hole off, leaving only a tiny hole with dynamite still in it that didn't explode.
So unless the miners clean very thoroughly with a hose, and wash down the face, and wash
it down real good, and look for those places bootleg holes could be overlooked. If you find one,
you're supposed to mark it with yellow chalk around the hole, so the next guy that comes
along will know there's a bootleg in there. That hole has to be loaded up with a quarter or
half stick of dynamite, and try to explode it, to get rid of it. It's a dangerous thing; it's a real
widow-maker!

In the next photo of a miner who's at a rock with a jackhammer, the jackhammer was
then called a J-55 or J-45, or a plugger. They'd drill a hole in that rock about a foot and a half,

Hard rock mines in a glory hole somewhere in the Nickel Plate Mine. 5 Miners in the "Good Old '64 Stope": "Vaydo", Jack Trudgeon, Eli Tootis, Dan Luxon, and Sig Floodstrom. The darker areas on the walls of the mine shaft of the stope indicate that the rock is rich in gold ore.

Reg French photo.

and put a half a stick of dynamite in it, and put it down, so that it would make 'Texas pea gravel' out of it. There are five guys working in the picture. The cavern underneath is called a 'stope'. They opened it up, meaning they drilled and blasted it, and you can almost see the ore body by the color in the picture. They'd open it up, and if there was solid rock in there, some of these stopes would go for about 300 or 400 feet. You needed solid rock on the top. There would be a presence of iron-iced porphyry that would be in there.

We had three locis, plus a battery driven one, that went to the Bulldog project. The big loci was the king of the lot, a very powerful machine, 440 volts; all the locis ran on 440. Later they brought up another loci, it was a much lighter one, I don't know just where they picked that up, but none of us ever liked that thing. It would never stay on the track because it was too light. It would spin its wheels and jump the track, and we were continually putting down the track jumpers to get it back on the railway tracks again.

There was an accident with a smaller loci. It jumped the track, and the track gave out from underneath it, and down she went! We got the big loci in behind and we managed to keep it on the rails with jacks, and we managed to get the small one back on the track, and repaired the whole place. The mine, at least the haulage part, was shut down about one day.

Outside of the mine portal was the cookhouse and the "dry". This was where the guys

Nickel Plate miners Sandy Brent, Charlie Pitt and Andy Kosley.
Sandy Brent photo

who came out of the mine would go in, shower, comb their hair, get all cleaned up, put on their day clothes, and then start off down the track to the bunkhouse, which was approximately one quarter of a mile from that site. Down the track, the next building was kind of a grease shed, and the next one was a warehouse. The taller part of the building was the main office for the mine. The building up the hill was the old 'hot air' bunkhouse. There was also a dynamite magazine for the mine. A little cart would be pushed back there and loaded up with the dynamite for going into the mine.

There were lights on all the time going down the Dickson Incline. The lights are also used as signals, blinking certain amounts on and off, to the hoistmen. If you wanted to come up you'd give them a three and one with the lights. If you wanted to go down and they were hauling too much you'd have to walk down. If you had to walk down, say to the 900 foot level, or down to the 1500 foot level, by the time you walked 1500 steps your legs really knew about it.

Joe and Andy Dzeures in the Dickson incline, 1938. At this time there were 1500 steps leading to the bottom of the incline. The Dickson incline was started in 1912. Gomer P. Jones received instructions to go ahead with a major development project, to sink a permanent shaft from the floor of No. 4 tunnel to a depth of 3,000 feet, reaching all of the ore-bodies and designed to remain as a great highway for the ores of the Nickel Plate mine for many years. The shaft was named the "Dickson Incline", after Mr. Dickson, vice-president of the company. The incline was sunk at an angle of 30 degrees from the horizontal. It was estimated that the shaft would have payable ore above it continuously for 1,100 feet. The shaft is 9 feet high and 18 feet wide and carries double tracks of 36-inch gauge. The Dickson Incline was first sunk to the 860-foot level, but later was continued to the 1,500-foot level. At the collar of the shaft, some 800 feet in from the portal of No. 4 adit, a Canadian Rand double-drum air-operated hoist was installed. In order to load the trains, an ore-pocket was cut, having a capacity for 50 tons of ore. The Dickson Incline was one of the mine's main shafts for the hoisting of ore.   Joe Bromley photo

Some of the Nickel Plate Mine crew coming off shift at the mine during the 1950's. L-R Charlie Wheeler, maintenance mechanic, Leroy Grainger, hoistman, Pete Wheeler, miner, Jim Camarta, miner, man behind not identified, Jim Wheeler, machine operator, Dolly wheeler, Jim's wife, and Howard Graham, miner.   Wheeler family photo

There were two skips in the Dickson Incline and the hoistmen would run both skips at the same time.

Lenard Wheeler was one year old when his family moved from Manitoba to the Central Hotel in Keremeos Centre. The year was 1926. His father, Charlie Wheeler, was born in England in 1893 and came to Canada in 1911, first settling in Manitoba. Eventually the family operated a dairy with twenty four milk cows, in conjunction with a twelve acre orchard in Keremeos. The milk was fed to pigs or just dumped, and the cream, worth about a dollar a gallon, was sent to Penticton by Greyhound bus. Charlie Wheeler sold the dairy cattle in 1939, shortly before the war broke out. The cattle were taken by truck down to a dairy farm in Wenatchee. Unfortunately, the prices paid for dairy products increased dramatically during the war. Lenard had to quit school the year he passed into grade nine to help his father on the farm. "I helped dad till the day he sold the cattle in '39. We sat down, and he said, "Well Son, that's it. No more hard work!" He went up to the Nickel Plate Mine and started work in 1940 just after the new year."

"My father started mining at first, and then they had him moving machinery in the mine such as tuggers, scrapers, and drilling equipment. He also did blasting. If there was a big rock caught in the grizzley he would drill the rock and blow it. Sometimes he used so much powder in the rock that he would blow the grizzley and everything out. That's why they called him Blaster Charlie! A grizzley is made of heavy steel, usually railway track rails, and is used to regulate the size of the chunks of ore that are dropped into the ore cars. My dad lived in the bunkhouse at the mine and came home on long change weekends, which were three days. The second weekend would be a short change, that would be one day off, and he wouldn't come down."

Lenard spoke also of his own association with the mine. "I went up in 1940, I was too young to go underground. I was only fifteen, so I worked in the carpentry shop. The only time I got to go underground was to measure timbers for the stopes and shafts. Then the carpenter and I would go back up, make them all up and send them down. From 1941 to 1943 my brother Pete and I worked in a mercury mine in Fort St. John. I started working for the mine again in the 1950's, but this time I worked in the mill. My job was to load five wheelbarrows of pebbles, hard lumps of rock from the mine, into the tube mill. The hole in the tube mill was about eight inches in diameter and constantly going around. You had to throw the chunks of ore in there. Mind you, I lost quite a few finger nails!

Mr. Biggs, the superintendent, asked me if I would like to learn how to operate the tram line. All through the summer I would take the place of the regular guys when they took their holidays or days off. When I ran the tram from the ore bin at the top, I stayed in the bunkhouse at the mine. When the regular man came back he let me down to the Central Station, in an ore car, to run the skip from Central to the mill. Every night the skip operator at the ore bin would bring me back up, and I would catch the train to the bunkhouse. In the wintertime, after the fellows had finished their holidays, I had a job as mine mechanic underground, servicing and fixing tuggers, scrapers, and mucking machines. I was there three years. My dad and my two brothers also worked in the mine, and my brother Jim, and his wife Dolly, lived up at Nickel Plate.

Jim Wheeler had the job of driving a Dodge 4x4 truck delivering coal and groceries around the mine site. He also had the job of cleaning the snow from the skating rink, which in summer was the tennis courts. The coal came up on freight cars that were pulled behind the ore cars on the tram line. These cars were about twelve feet long, fitted with side racks, and then filled with coal. When the cars reached the top it was pulled onto a side track by a tugger, and moved to the railroad, where it was hauled behind the electric locomotive to the mine and townsite. When the roads became better, trucks hauled the coal up to the top. Jim finally got a job underground using a mucking machine loading ore cars. He got an hourly rate plus a bonus. The mucking machine ran on tracks and threw the ore over the machine into the ore cars on a side track beside the mucking machine, which were then hauled to the surface and taken to the tram line.

Barney started mining also, and as soon as he could, started working on a mucking machine. The miners liked working on the mucking machines because of the bonuses that were paid. Barney was at the Nickel Plate Mine two years and then went to Burns Lake with our brother Pete.

My dad, Charlie Wheeler, worked at the mine till it closed, in fact he bought a Nickel Plate house from up there. When he left he wanted a radio and a bed, the same old cot he had all the time he was up there. He used that bed till he moved to a home in Summerland. He was still sharp when he died at 94 years. My brothers, Pete, Jim, and Barney are gone now also."

Stan Jones was one of the youngest and last people to work at the Nickel Plate Mine. His father, Cecil Jones, was the head electrician at the mine, which no doubt helped him find employment with the mining complex. Stan worked for the mine during school holidays, washing windows and pulling weeds around the mill, which gave him credibility as an employee.

"When I started working at the mine I was too young to go underground. I was seventeen years old and you had to be eighteen to go underground. The other kids fibbed and got to work underground as muckers loading ore cars or working with the powdermen. My dad worked for the mine so our family records were in the office. Working underground you often got a bonus for production, and that was where they made good money.

I worked on the tram line as a brakie, pulling ore cars to the ore bin, unloading them, and returning the empty ore cars to the mine. When I was old enough to go underground, I worked with an outfit called Connors Diamond Drilling, looking for high grade ore pockets to keep the mine open. Apparently we didn't find any!

I stayed in the bunkhouse at the mine, but anytime we wanted to skate or play ball it was usually down in Hedley. At these times I usually stole a broom and rode the tram line rails down to Hedley, and it didn't take long to get there! In the winter time there was a groove in the snow made by the wheels of the tram car. From the ore car down you were doing a pretty good clip; when you hit the trestle you were really moving! There wasn't any snow on the trestle to slow you down. I wore out a few pair of gum boots on that tram line. I usually stayed at my parent's place in Hedley, but then I had to hike back up the mountain in the morning. It was the brakie's job to turn the switches on at the generators at the mine, which produced the 550 volts DC for the locis.

I started working at the mine in the summer of 1953, and stayed through the year till it closed in 1954. Being one of the youngest, I was the first laid off. I moved the Britannia Beach gold and copper mine at Squamish, and stayed at the townsite at the top of the mountain. It was a set up similar to the mine at Hedley." Stan is now retired and lives in Penticton.

13. THE LOCI AND THE RAIL LINES

Bill Orser, a long time resident and miner, recalls his ominous arrival at Hedley in 1923, when he was three years old. His parents were penniless, so they had walked much of the way from Princeton to Hedley, with the exception of the last four miles, when they finally got a ride. Bill relates that they had no money at all; his dad just had a guitar - of all things! They were fortunate on their arrival, in that a certain big negro fellow was running the kitchen in the hotel, and he had compassion on the poor family when Bill's dad walked in and said he had a wife and child waiting outside, exhausted from their long trek.

The Negro said: "Mister, you just go sit down and fill your face, and you make sure the Missus and the kid fill their faces, too! I'll foot the bill, you'll get a job."

Three days later Bill's father, Sam Orser, did get a job, and in the meantime, the Negro kindly kept them in food. They lodged at the hotel, while Sam worked double shifts. On the fourth day his employers asked him if he'd like to go up to the Nickel Plate to live. This was a real blessing to the family, as Bill says, "I'll tell you it was really something to have some-

The electric tramway, from its terminus at the ore-bin station to the portal of No. 3 tunnel, is about two miles in length. Because of tunnels along the mountain slope several hundred feet beyond the portal, a switchback allowed the ore-trains to be backed from there into No. 3 tunnel. Tracks of both electric railroad and gravity tramways were laid to a 36-inch gauge, and 20-pound steel rails were used on both tramways. The ore-cars used on the electric tramway had a capacity of about 2 tons of ore.
                                                              Pat Wright photo collection

An ore train and passengers heading for the ore bin at the top of the tram line.

Joe Bromley photo

thing, just a shack, but it was a home of our own, a place where we could stay. There was no rent. Dad had a job; things looked pretty nice!"

The family had been at Silverton Mines before they came to Hedley. Sam was laid off work there, and they had stayed around for some time waiting for the mine to re-open, but it never did. Bill's parents had spent all the money they had, even resorting to selling their furniture to buy food. However, they did manage to save enough to make the trip through to Vancouver, but to no avail. Then Sam heard that Hedley was hiring men for the mine there. Bill and his mother had no place to stay in Vancouver alone, so they all left together. Their carefully saved money finally ran out at Princeton, hence the reason for the days of arduous walking for the young family from Princeton to Hedley, a distance of some forty miles.

Bill Orser himself, took his first job in the mine several years later when he was sixteen years old. "My first job was mucking," Bill says. "Shovelling the ore into a wheelbarrow, wheeling it along a plank, and putting it down a chute, all done by the light of a carbide lamp; this was in 1935. I went drilling, and finally worked my way up to be a miner at last. It took a long time, because you were working with Cousin Jacks, and they didn't want to teach the young fellows too much."

Bill explains that "Cousin Jacks" were men who had worked in the mines in England. They had left their wives in the early 1900's, come to this country, and obtained jobs in the mines. They were mostly muckers, or powdermen; they hardly ever went as miners. They

Bill Orser, brakeman, Jack Evans, motorman, hauling ore from the Nickel Plate Mine in 1935 to the top of the tram where the ore was transferred to the tram car for a trip to the Daly Concentrator 4000 feet below. The brakeman used mitts rather than gloves to turn the brake wheel, which pushed a steel plate against the rails for friction. Gloves wore out too quickly.   Bill Orser photo

seemed to live over here and send money back to England, to their wives and families, who they hadn't seen for twenty or twenty-five years, and never really did expect to see them again.

Bill Orser was for a time brakeman on the ore trains. "We used to come down with the ore train to the ore bin where we had to dump the cars. The idea was that the train would back down from the old powder magazine, and the brakie, who was me, would jump off the train and run ahead, reach in and get the coupling pin, pull the pin out and throw it to one side. Then I would follow the car in to where the dump was, and push as hard as I could and get 'er in there. Then the ore car would tip, and come back out and the return spring would release on the track, and the ore car came back out on another set of tracks. The spring was on the rails, so the car would hit the spring and bounce it back out. The dump looked like one of the cars in itself, the rails came right into it, and the rails were hinged. We would take one of our ore cars and run it in there, and when it hit in there with a bang, it would unhinge and tip right over, and dump the ore right into the ore bin. When it came out, there was a spring that made it automatic, we didn't have to do nothin'. The whole thing with the weight of the ore in it would flip and tip the ore, and away we'd go. We did these one at a time. When I was looking out of the ore bin, I'd watch that tram go up and down, it was 4000 feet down there below me.

This is a loci hauling a load of ore cars to the ore bin at the top of the tram line. The two electric "locis" ran on 440 volts D.C. power. The crew consisted of a motorman and a brakeman. Each car held about two tons of ore. In later years the locis were changed to 550 volts D.C. electricity.

Cecil Jones photo

We were coming down the skip one time, we'd just gotten off shift. We heard the whistle go, and old Yorkie said, "Aha, there's an accident in the mill". We got down further and we saw they were taking this man, I don't know his name, but he had fallen down about 40 feet, crossways onto timber, off one beam, and crossways onto another one, and then another beam, until he hit the floor. Of course he was dead when he hit the floor. Kind of spoiled our day in Hedley!

Gomer Jones used to come up from Hedley out of the main office, and he'd be with a fellow by the name of Jack Biggs, who was the project engineer. Gomer always seemed to get a great kick out of it, and he always talked to me. He was somewhat like a grandfather to me in a way. He had to marvel how I could get in between those ore cars and back out again . He'd say, "You sure want to take it easy, Billy! It's pretty deadly between those cars!"

Bill Orser recalled a story which he had heard as a youngster on Nickel Plate mountain. Jack McClafferty, one of the workers, had got married while working at the Nickel Plate and moved into a house in Bogus Town. Camaraderie among the men in this small isolated mountain top village was strong. The boys were excited and decided to plan a chivaree for him. Through the day each shift had gone into the main office and donated one shift, a day's pay, to Jack and his bride, which, at about $5.00 a day, and with three hundred men working, accumulated to quite a gift, especially for those days!

A trainload of ore and passengers heading for the tram and a trip in an ore car down to Hedley. The man standing is Alex Shaak, the mine superintendent. The date is March 5, 1945.

Sid Dannhauer photo

Jack treated the dayshift royally at 4:30 in the afternoon. The afternoon shift was also treated royally with drinks and lots of laughter, but when the graveyard crew came off shift they thought they would liven things up a bit. A couple of the boys took a few sticks of powder from the mine and set them in different places, quite a bit back from the houses so they wouldn't break any windows, and then let them go.

It was terrifying for everyone up there to hear these four sticks go off. Everyone rushed forward making a lot of noise until they got up to Jack's door. The door opened up slowly, about four inches wide and the barrel of a 30:30 rifle stuck out.

"O.K., you sons-of-bitches, get out of here!", said the voice behind the gun. The plan had turned sour! The men were all so mad, they promptly went and told the timekeeper, who was still around, to strike their names from the $5.00 donation list. McClafferty said goodbye to a considerable amount of money that night!

The humorous part of the story concerns Mrs. Vaydo, who lived next door to the Orsers, and vividly and colorfully told Mrs. Orser about the prank. "The first shot I wake up; the second shot I out and onto my knees; the third shot I start praying; and the fourth shot I say, 'My God and Saint Mary, here I come!"

Bill Orser and his mother and father, Sam, hold the speed record for a ride down the tramline. Bill, who now resides in Okanagan Falls, recounted how it happened.

The electric loci transferred to the rear of the ore car train and moved the cars into the ore bin. The cars were uncoupled, and pushed into the ore bin where they were automatically dumped, using a spring loaded dumping mechanism.   Roffel/Budd photo

"I got laid off, it was in the fall of the year and we were headed for Sicamous. We got down the tram as far as Central, and lots of niceties were being exchanged while Bill Edgerton was switching the cable. He never mentioned that two of the three brakes had been taken off the Central drum, or Dad would never have let us go on. Anyway, we started down, everything was fine, when all at once the cable went slack. We were sitting on a skip filled with ore, about 5 tons, with a freight car in front of it! We were going down that hill pretty good! We even saw leaves and rocks being sucked up from the vacuum created by our skip, we were going that fast!

Edgerton had put his shoulder against the bullwheel and pressed as hard as he possibly could. It was actually the heat and expansion of that brake that slowed us down, within 200 feet of the dump. He, Edgerton, stopped her up within 75 or 100 feet of where she dumped! I had thrown my guitar off; I often look back at it as very ironic, here I was going down the tramline, and it looked as if we were going right through, and I had my harp all ready!

They blew the whistle in Hedley signalling that there was an accident. People looked up and they could see that skip coming down. They said it was drawing a line it was coming down that fast, and then it slowed right up. I was all ready to jump into the sand. There was a big pile of sand there, but then I felt the skip starting to slow up, and I figured we could stand all right.

From the ore-bin, the upper terminus of the gravity tramway, a wonderful view can be had of the snowy peaks of the Cascade Range away to the west and north-west, as far as the eye can see. To the south and south-east lie the jagged peaks of the Ashnola Mountains. In the immediate foreground one looks down into the narrow trench-like valley of the Similkameen, some 3,700 feet lower at river-level.   Cecil Jones photo

Mother stood up fine for about an hour and a half afterwards, until we were over at Mrs. Joe Bromley's place, where she fainted from shock. Dad was white, and I wasn't doing too well myself. My knees were pretty shaky. According to Edgerton and a few others, they say the Orser family made a speed record going down from Central to Hedley on the skip! Believe me it was some experience!"

The Orser family got over the initial shock of their experience and went about their daily lives. The incident, unfortunately, was not as simple for Dave Edgerton who had been running the tram at the time. The potential disaster, which could have happened, played on his mind until it caused him to have a nervous breakdown.

One of the problems of clearing the rail line with the cleat-track tractor was that the tracks did not fit inside or outside of the locomotive rails. Consequently one track was between the rails and the other on the outside with about 4 to 6 inches of the track hanging over the railroad ties, not a problem on flat ground, but a bit unnerving on the trestles.   Arthur Irwin Photo

Snow removal was a problem in the wintertime at the 5500 foot level in elevation. The rail line was the only route that was kept clear of snow at all times. Roads, until autos became popular, were left unplowed. Walkway staircases had a canopy to keep the snow off.   Arthur Irwin photo

14. KELOWNA EXPLORATION CO. LTD.

For anyone who has not lived through or worked during the depression of the 1930's it is difficult to fathom the living conditions of those times. What is equally as difficult to comprehend is the attitude and actions of the employers in that time period. Money was scarce, the market was tight, and labour was cheap. Kelowna Exploration Co. Ltd. was possibly the same as any other company. It is not the intent to ridicule the company or company policy, or the employees. The intent is to reconstruct the events of that time as it was. The best source of information available about the people and living conditions is the people who lived there.

When the mine was reactivated by Kelowna Exploration Company, there were some changes made. The original flume from Twenty Mile Creek was abandoned, and a new dam and metal pipeline was installed. A reservoir or holding tank, which was still in use until recently, was built to feed the mill and supply domestic water to the town. A water line went across the side hill onto the crusher floor down through the mill, and took the waste down to the tailings pond. This water system was used until a flood in 1972 destroyed the dam and intake, the town of Hedley presently has a well and pump, which utilizes the reservoir built in 1934.

Hauling freight to the Nickel Plate Mine, recently taken over by Kelowna Exploration Co. Ltd. in 1932. The white board gave the times for the one way traffic to go up or down. Harry Allison is the driver of the truck.

Dave Innis Photo

Mine machinery direct from England going to the recently formed Kelowna Exploration Company Ltd. Mine via the old Nickel Plate Road. Dave and Hyl Innis began their freighting career hauling supplies to the mines in the Cariboo with their trademark grey Percheron teams. Later, Dave operated the Palace Livery in Upper Keremeos, then moved to Keremeos where his business changed from horses to trucks. Hyl Innis operated J.H. Innis Garage Trucking in Hedley.

Dave Innis Photo

Mr. Jack Biggs, who was mechanical and electrical foreman of the Nickel Plate Mine during the time when Kelowna Exploration Co. Ltd. was in operation, described briefly the workings of the hoists in the mine. Jack related that the Dickson hoist was located at the head of the Dickson Incline at the same level as number 4 adit. The cables for the hoist ran upward for about 125 feet, to where the cable ran on two sheaves which allowed the ore cars to dump into a pocket. The "pocket" was a cavity above the level of the surface train which allowed the ore, or "muck", to be loaded into the ore cars by means of gravity.

The hoists were located in the "hoist room". Off that room was the "compressor room". The compressor room contained two compressors and a motor generator set that supplied power to the electric tram. The locomotives, one a 10 ton and the other a 7 1/2 ton, ran on 550 volts D.C., changed from the original 440 volts D.C. The electricity came from the West Kootenay Light and Power Co. at 2300 volts. A large electric motor ran a direct current generator which supplied the 550 volts D.C., which was sent outside and ran the trolley. There was one generator in the mine, and a standby unit at the upper ore bin. The room also contained two compressors that produced about 2000 cubic feet of air each, supplying the miners with air for their drills.

The Dickson Incline was at a 30 degree incline and about 1800 feet in length to the 1500 foot level. Levels in the mine were calculated from a base level of where the top of the incline

Switchbacks on the Old Nickel Plate Road leading to the mine. Harry Allison is leaning on the first truck to have this photo taken. This switchback road branching off the Green Mountain Road was laid out by L.A. Clark, who later started the Green Mountain Stopping House. Four and Six-horse teams often had difficulty navigating freight wagons around these sharp curves.

Dave Innis Photo

started; elevation usually denoted the elevation as calculated from sea level. Off the incline would be various stations where the actual mining took place.

Communication within the mine was through electric bells. The tender, when the ore car was filled, would juggle the switch and send a message through to the hoist room where the hoistman, by watching his gauges, could tell when the skip would be nearing the pocket ready to unload.

Howard Graham, at one point in his career at the Nickel Plate used to be a hoist man on the Dickson Incline. "I used to run that hoist. There was the compressor room just behind, and an air hoist just behind that, which was to be used in case of emergency. They'd wind that up once in a while, and she'd snort and blow water all over the floor. All the signals were electric. There were signals at every level. There were two drums in the hoistroom each with a signal bell. They had to have a different sounding bell so that you wouldn't get mixed up. One bell had a hacksaw cut in it so it would make another sound for one side.

Those guys down below used to give you a bad time. They would stop you down at a station, say the 800 foot level where the skips would stop side by side, because a level went off on each side of the shaft. Both operators would get on the bells but you would get used to it and they would never fool you."

~ 134 ~

G.A. Riddle Service Station with George Riddle serving gas to Hyl Innis. George Riddle had pre-empted land earlier near the Brent Ranch at Shingle Creek. George operated the Shell Station, however, both the Shell station and the Esso Imperial station belonged to Hyl Innis. Dave Innis, whose photo collection is used, was Hyl's son. Hyl and Dave Innis Sr. each had a freighting business.
                                                                                    Dave Innis photo

Miller Kirkpatrick running the hoist, and Mr. Turner, the mine manager, (extreme right), in the Dickson hoist room (main hoist shaft) for the mine.   Reg French photo

The bells, the telephone, and the generator used to keep you hopping, the biggest worry was whether you were going to put someone in the dump or not when you brought a car to the surface."

The Morning Incline was located below the Dickson Incline and at an angle that would allow it to tap the ore located in the Morning claims. In order to tunnel to the Morning claim the shaft which was at a 55 degree slope had to traverse one apex of the Mascot Fraction. The Hedley Mascot Mining Co. was later reimbursed for this trespass, which had been accomplished while Duncan Woods still held title to the claim. The hoist on the Morning Incline was identical to the hoist system on the Dickson Incline.

In the last year of the mine's operation, 90,000 tons of ore which yielded 36,303 ounces of gold was taken from the mine. The majority of the ore which was from the Morning claim would start at 4100 ft. elevation on a skip, up the 55 degree Morning Incline, to a pocket at the top of the steep tunnel. From there the ore was transferred to a skip on the Dickson Incline where it was again dumped into a pocket. From the Dickson pocket, the top of the Dickson Incline, which was Number 4 adit at 5600 ft. elevation, the ore was transferred into the ore bin, the elevation of which was about 5000 ft. From the ore bin the ore was transferred into cars on the tramline, and eventually reached the ore bin above the mill. From the mill ore bin the ore went through stamps and the cyanide process and emerged as concentrate, which was loaded into box cars on the Great Northern Railroad and sent to Tacoma to be melted into gold bricks.

Miller Kirkpatrick and Bill Richardson in the Dickson hoist room of the Kelowna Exploration Co. Ltd. Mine.
Reg French Photo

One of the trucking firms that hauled a lot of lumber and machinery to the Mascot and Nickel Plate mines, was Innis Trucking. Hylard Innis, or "Hyl" as he was known, had a garage and a trucking firm. Hyl started in Hedley, with a livery stable at first, and then as automobiles and trucks replaced horses, he replaced his livery with a garage. Hyl ran the Esso garage which had service bays for repairing or servicing vehicles. He also owned the Shell gasoline station, which sold only gas and lubricants. George Riddle, for whom Riddle Creek, a tributary of Shingle Creek, is named, ran the Shell station for Hyl. Hyls' brother Dave who was one of the first freighters to the Nickel Plate had much the same type of business in Keremeos.

Hyl looked after the service station and office, and had drivers for his trucks. One of the drivers was Johnny Allison; another person who drove for him was Sandy Brent. Until 1937 freight or supplies from Hedley went 53 miles around via the Green Mountain - Nickel Plate road to reach the mine. The road, with its many switchbacks and sand, had times posted at the bottom for uphill traffic and downhill traffic. Sandy recollected a time when he and Johnny were hauling lumber to the mine site. Johnny's load began to slip backwards; Sandy who was in the truck behind him noticed this. Not wanting to have to help him reload his cargo, Sandy stopped his truck, ran ahead and jumped on John's truck. Apparently Johnny had his truck in the lowest gear and full throttle - vehicles usually had a throttle and an accelerator then - but was travelling very slowly. Sandy climbed over the lumber and pound-

Still in Dickson hoist room. (Showing ladder up middle of stope). Earl Edwards, Trig Anderson, Mr. Turner the superintendent or mine manager, and on the motor tram car, Bill Cartwright.

Reg French Photo

ed on the truck cab roof, trucks also had roofs one could pound on at that time! Johnny apparently ejected from the cab with a look of amazement, wondering what had, or was going to attack him. The men shifted the load and continued on their way.

The Nickel Plate road was maintained periodically. A four horse scraper was used up until the 1930's to grade the road. The team would haul the grader up the hill with the blade disengaged. At the top the blade was lowered by means of two large wheels in front of the driver and the outfit would "drag" the road down. The same process was repeated again, and so the other side of the road was graded. After this process the road was hopefully devoid of potholes, washboard, or washouts for a time. In the late 30's early 40's the road was graded by means of a tractor and scraper.

Sandy started work for the mining company, but soon discovered that it was much more profitable to be a contract miner than a regular mine employee. He teamed up with various partners and tunnelled many of the drifts in the mine. As a contract miner one was paid $8.00 per foot, but you had to supply your own powder. The air machines and blasting caps were supplied by the company. As a contract miner you carried your own steel and drill oil in and out. The company had a policy that miners must blast every shift, so that meant packing in powder also. Another company policy was that the miners were not allowed to "burn" a raise, which meant drilling a number of holes close together, loading it with dynamite and then

The compressor room on the Dickson Incline. The electric motor in the foreground drove a generator which created 550 volts D.C. for the electric trains.   Reg French Photo

working all the ore to this central cavity. The conventional method of mining is to drill a draw cut, or a diamond cut, pack the holes with powder, blast them, and "muck" the ore out.

As a single man employed by the Kelowna Exploration Co. in the late 30's, early 40's, wages were about $2.50 per day, board was $1.00 per day, which entitled one to a room in the bunkhouse and meals in the cookhouse. Accommodation was adequate and the meals were good. As a "contract" miner, if you worked hard, you earned around $5.50 to $6.00 per shift, however, the board was increased to $2.00 per day.

Kelowna Exploration Mines and Kelowna Mines were the same company. They dropped the "Exploration" when they were satisfied they weren't going to find any more ore. At one time there were 42 drilling machines working, operating in the mine, two were for the mill and the rest were for exploration.

Joe Bromley remembers: "When the Kelowna Exploration Co. started up the mill again in 1934, they hired an American man who came down from Alaska, by the name of Dustin. They put him in charge of the power plant. It used to be that when they turned the water in to the flume, they could only turn it in slowly. You'd open the head gate and wait for the flood of water to go down, then open it another inch or so. The men that worked had been on there for years, and they knew how to handle that thing. They had to shut the water off at high water time, to clean the flume from logs in there. They did that, and then when they turned it on, and started it up again, this Dustin came along and asked why they didn't open it right up.

The 4300 foot station on the Morning Incline.   Reg French Photo

They told him they couldn't, that it would wash everything out down there when the water hits the forebay; it would take it right out. The flume came down and there was a big wooden box built in the water. The box was called the forebay. The water would turn, then go down the penstock and through the turbine. Anyways, he kept telling them to open it up. They argued with him. He gave the orders to open it up. He said he would be responsible. They opened it up and the water came down in a rush and hit the forebay and took that forebay right out! Then it started out and washed a gorge out, and went through the windows and filled in where the turbines were located. Filled it all full of gravel! They had a channel made out to the river about a quarter of a mile long, about 50 feet wide, with sloped banks. It was a terrible mess down there. The channel was all filled with gravel and had to be all cleaned out again by a bulldozer.

They had to rebuild the whole thing, and get timbers dragged up. They sent me down from the mine with a crew and we worked on the night shift and had to get a whole bunch of timber dragged up there. We pulled the big timber up with a winch. There was some of that forebay still hanging in there. I worked on the night shift tearing that out. We used carbide lamps as there wasn't any power down there. Dustin got fired I think; he left shortly after that. We had to get a bulldozer in there. Water was carried by a flume 9 feet wide from the dam to the powerhouse, three miles down the valley. At the end of the flume the water ran into the forebay and then down a penstock and through the turbine that ran the generators to produce electric power. This was what was all washed out.

When they got through, the river had risen and they couldn't get back. The bulldozer had

This was the Nickel Plate Mine Rescue Team in June of 1952. The men are L-R: Ray Richards, Bill Richcun, Tony Triggs, Collin Brown, and Ted Peck. Kneeling are Dibbs Williams, Len James and the mine manager, George Mills.
Doug Cox photo collection

forded the river to get over there. So they had to go back, it was a big heavy bulldozer, and they ran it clear across the dam. It was a concrete dam, and they had these logs that if the water was high, they could pull by winch. The first pier broke right in half, it opened up in the center. They hadn't got going very long when that happened. So they put me there with a bunch of men on night shift trying to drill with hand drills. They were going to blast it out and build a new one, but that thing was full of steel and we had an awful time drilling and blasting it. They brought a guy in from Vancouver to look at it. I had blasted the pier at 6 a.m. when all these big shots all came down to watch. They went across the pier on the other side and they were all standing on top of the pier when it cracked open. The men were standing around on it and it opened right up. Scared the wits out of them! They took some cement off the top, built a form about two feet out all the way around it, and they got by with that. There was still a crack in the other one.

They got the dam going again and then high water came. They had men pulling the stop blocks but the whole dam collapsed. The piers collapsed, and they had to run for their lives to get off the dam. It was washed right out and was no good anymore. So then they hooked on to the Kootenay Power. That was in 1936.

They had started up the mill again in 1934. They spent $30,000 fixing up the flume down to the power plant. It may have been in 1935 when the power plant was washed out, or 1936, I can't remember. The first year it was the power house, the second year it was the flume.

In the early 1950's it was necessary to bring in equipment to dredge out and dam Nickel Plate Lake (formerly Stray Horse) to ensure a steadier supply of water for the mill at Hedley.   Biggs photo

There were two small air compressors in the mine. If they needed more air, these compressors kicked in and were supplemental. If the pressure went up they would go off again. The hoistman on my shift was Andy Dzures. The operator had to look sideways while he was running the two hoists to see the air gauge. When the air pressure went down he had to shut off the hoists and go over and start the compressor, and then go back and watch for a while. When the air came up to the right 94 pounds per square inch, he had to go back and shut it off again. It was too much for one hoistman. There was a telephone there, and somebody was forever phoning the hoistroom, either the office outside, or the mine. There were telephones on all the levels. He just about went crazy, trying to answer the telephone and run the hoist; there would likely be an accident, and then of course, he'd be blamed for it. Finally they rigged up an automatic device here. If they had the gauge out at the front where the hoistman could have watched it, it would have been better. Jack Biggs was in charge of the machinery part and the tramway.

There was a train that went off at the ore bin. They had let the grass grow up on the electric tramway. There was an awful rainstorm and grass got over the rails. Two brothers, motorman and brakeman Charlie and Alex Saunders had trouble going down, and they had an awful time getting the loci and cars stopped. They had loose sand on the tracks for traction for the motor. When they got back, they were afraid to go back down again. They used to trade motors halfway from the mine to the ore bins. They had one big loci and one small

This was the water gate being installed at Nickel Plate lake. The dam raised the level of the lake and created a greater reservoir and supply of water for the mill. At one time steam engines were used to pump the water out of the lake.   Biggs photo

motor. They used the small one to haul the ore from out of the tunnels. They would pass with the empty cars coming up, and the big one would take over the train. They used the big motor at the lower end because it had more power. When they got back up to the passing place, they phoned the office to speak to the superintendent, and told him they were afraid to go out anymore because the rails were just like grease with the grass over them. He just said to go ahead and use the sand. So they went down with a load and couldn't hold it. They had hand brakes on one car in the middle of 18 cars. The brakeman put on the hand brake as tight as he could, but the wheels would just slide out. Before they got to the ore bin, they jumped off, the ore train was running away. The motor and 2 or 3 cars just went right off the ore bin. It was pretty high there, sixty feet high I guess. They had quite a job to get the motor and cars back up to the tracks again. Bill Pearce, a mechanic at the mine, got his leg broken accidentally while getting the motor and cars back up.

The pipeline that carried the compressed air from Hedley up to the mine passed though about 1/4 mile north of the ore bin. They closed down on every second Sunday, from Saturday evening at six o'clock until Monday morning at seven o'clock. I was the shift boss (Joe Bromley) in charge at the mine. The superintendent and the engineer and all other bosses were away. It was my turn, and Jack Biggs sent three fellows up with an acetylene welding outfit. They took it up with a pickup truck and got around as far as they could. They couldn't get the truck down as far as the ore bin. They packed the oxygen bottle and other assembly down. It was dry weather and Biggs sent these two fellows along with two fire extinguishers, to watch in case they started a fire. The welder got started and they went away and left him. He was welding in there, and all of a sudden he felt his back getting hot, and he looked around and the bush was all on fire. He started running up to the ore bin. Theo Prest, who was operator at the ore bin, took his automobile part way down to the ore bin. There was a road down there and he could get pretty close. He came rushing up to the mine and wanted to get some shovels. He told me that there was a fire burning down there in the timber below the ore bin. I asked if they needed some help. There were two fellows who had gone down from the mine, Bill Triplett and Doug Kirkpatrick. He took some shovels from the warehouse, and I got worried about it, it was so dry. I finally got Jack Biggs on the phone and told him about it. He said not to take any chances. So I rushed down to the Community Hall and there was a bunch down in the basement playing pool. I told them to run for the ore bin. Everybody I saw, I told them to run for the ore bin. I went to the warehouse and there wasn't any more shovels in the warehouse.

I had some men working in the tunnel then, and they had some shovels. I told them to take a saw and a couple of axes and they rushed down there as hard as they could run. There was a young fellow from the survey crew, and I asked him to stay in the office to look after the phone. Pretty soon they phoned up and asked me to get some water down. They had a tank of water there, but it was all used up. The fire had burned up to the ore bin, and all around it. Some of the timbers of the track started burning, also the trestle underside. Mr. & Mrs. Theo Prest. used to pull water up in the tank, from Hedley. I ran down and got the big motor out of the shed. I came down to the cookhouse and there happened to be a big oil drum down there, open in one end. I got that out and the Chinese cooks helped me fill the drum. It took me quite a while to fill it from the tap in the cookhouse. I ran up to the bunkhouse and the other motorman was up there. I asked him to get the small motor out and try and find another drum. I went down to the ore bin with the drum of water on the big motor. It was lucky I got there with it; the ore bin was still burning. There was a pole with electric wires on it all on fire. A young fellow took a bucket of water and he shimmied up the pole and poured the bucket of water over the top, onto the fire. He got the fire out! The generators that run the electric motors, were in the bottom of the ore bin. If the ore bin had burned up, all those generators that run the electric motors up there would have burned.

In the meantime I guess, Biggs got alarmed, or maybe Theo Prest phoned him up. He wanted to know what was going on. They told him that the fire was right to the ore bin. So they started the tram running and got a bunch of men and tools up from Hedley. It was a battle to get that fire under control. It partly burned the trestle. That was in 1942-43."

Bill Orser started working at the Nickel Plate Mine in 1933. He recalled Billingsley coming to the mine: "Billingsley and his daughter I assume came up from South America. Kelowna Explorations had a mine called the Little Bear Mine in South America. They came up, and Miss Billingsley, being a young woman around a mine at that time, was something

Jack Biggs was a project engineer at the mine and later the Provincial Mine Inspector.

Biggs photo

for the men to really look at. She had a wonderful shape! She came in there and she had on breeches, leather high topped shoes, a shoulder strap down to her waist, a hard hat on, and a little rock pick in her hand. She called in to the lamp house at the dry for a lamp. The lamp man that was on at that time was a man from England, whom we called Cousin Jack. He said "Ma'am, what would you be wantin the light for?"

She said "I'm going underground".

"Oh no yer not", he says, "You'll upset the whole mine!"

She said, "I don't see why, I've been in all the other mines."

He said, "It's a going thing now, here, that if a woman enters a mine, it'll bring on bad luck, so therefore half the boys, from England at least, will walk out of the mine and will not go back in it."

"Aw" she said, "it's all garbage. Give me the lamp!" She had the authority and she got her lamp. She came down the mine and scared a lot of the 'Cousin Jacks' pretty bad. However, they were only off for a day or two, before they came back. They saw it was a whole lot of melarky, a woman had as much right in there as a man. She went along getting chip samples here and there. She did a very wonderful job of building a model of the Nickel Plate Mountain, complete with the mine in it. It was all built out of glass. The Dickson Incline was in red, and you could practically see right through it at different angles. It was a huge piece of glass, and she put it all together, and colored all the drifts and stopes, and the inclines.

These young men, Sandy Brent, Jim Jamieson, Dick Lowe and Buck Allison, posed for a photo in front of Mrs. Edmund's boarding house and cafe, Hedley, 1934. Hedley was booming at this time, as the Nickel Plate Mine had re-opened and the Mascot Mine was constructing buildings prior to its opening.   Ralph Overton photo

Everything in it was different colors. She had it all mapped out and built this thing. It was approximately four by four by four feet in size. She kept it pretty much to herself. I was lucky enough that I got in to see it at one time. Mr. Turner saw me come out and asked me what I was doing in there. I said that I just wanted to see what she had been doing. He said, "It's really none of your business, so you just stay out of here from now on." This was in the Kelowna Explorations office at the Nickel Plate Mine.

Billingsley was fairly old, and he came up mostly with his daughter. His daughter did the leg work in and around the mine. She would go down, and a fellow by the name of Perry Palumbo, a little Italian fellow, was the sample man. He was the man that would bring around the electric blasting caps or take fuse caps for blasting. That was his job, making

This was the assay building at the Daly Concentrator. The concentrated ore was sent to the smelter in Tacoma, Washington, where the arsenic pyrite which turned the gold black, could be removed. Some of the finer gold concentrate was sent to New Jersey, also in the U.S.A.   Jack Bottaro photo

them up. His other job was picking up the samples of rock from each station. He'd come up and he'd have probably 150 to 200 pounds of rock, bringing it out on a little car. That would go to the sample shack. Miss Billingsley would get a little sample of everything from Perry Palumbo, as well as going through different drifts and places on her own and getting samples of her own. She cut quite a figure there, and, although quite a few of the younger fellows tried to date her, she'd have nothing to do with anyone. She was quite a striking figure in her tall legging boots.

Her dad mostly sat down at the main office down in Hedley, he'd come up just once in a while. He'd be over at the cookhouse drinking coffee or sitting at the main office at the mine, and she'd be doing all the legwork going down and getting all the samples he wanted. She built that glass mountain, and I've often wondered where did the glass mountain go, I'd like to know?

I started at the mine in the fall of 1933, when I was fourteen and a half. I stopped working there in October, 1939. War had been declared, and I joined up."

Jack Bottaro was an energetic enthusiastic man when I contacted him about 60 years after he had started working at Hedley, and eventually with Kelowna Exploration Co. Ltd. in the assay office. Like many young men during the 1930's he was riding the rails looking for employment, or at least something to do.

Here are some of his reminicenses of Hedley. "I remember going up through the Coquihalla, this was mid-summer. We'd get off the front end of the train and pick berries till the caboose came along and then climb back on the same train again! We got to Brookmere, and we were in Brookmere overnight. That evening we had a softball game between the railroaders and the hobos. I can't remember who won and who didn't, but we enjoyed it. Some of the railroaders had some stale meat and dry bread and we were happy to have it. We finally made it down to Princeton. In Princeton I was met by one of the truckers that worked for Hyl Innis. The trucker was Johnny Allison, who was a great grandson of the 'Indian' Allison family, which the Allison passes were named after.

We got down to Hedley and I went to work for Hyl Innis. The Kelowna Exploration Company, which was an American interest that had bought up the Nickel Plate property and the concentrator, was getting ready, and in the next year they would get into operation. Consequently, there was quite a lot of activity around Hedley at the time. Anyone who had any skills at all could get work. The one problem of being a mechanic in Hedley, as opposed to elsewhere, was that in the winter there was no traffic, so there was no need for a mechanic. There wasn't a railroad operating in Hedley at that time. The steel was still there from Hedley on down, but not from Hedley to Princeton. The company needed a crew in their power plant to run the boilers, and since I'd had power plant experience, it was fine. On Christmas Day 1934, I went to work for the Kelowna Exploration Company, since Hyl didn't need me in the winter time. I went to work shoveling coal in the power plant.

There were three boilers together on one stack, and one boiler on its own stack. I worked with Billy Knowles. There were two firemen for the four boilers. The coal we were burning was slack coal that was hauled down from Blakeburn. You had to be very careful because if you put a hook or ash raker in it at the wrong time it would literally turn to molten glass, and there would be problems. Not only would it be tough to get out, but it was also difficult to maintain heat. 'Slack' coal had a lot of dirt and rock in it. Blakeburn coal was the best coal, some of it would even qualify as anthracite. The bituminous coal was a domestic coal, which was ideal for heating, for firing boilers, for maintaining heat in a home, or any purpose like that. It burned clean. The smoke was clean with not too much ash, and hardly any clinkers. As it got mixed with dirt or rock you started getting clinkers and you had problems. You had to be careful how you fired it, and that you had it fired evenly all around. The fire created steam. There was a steam engine at the far end of the powerhouse; it was a Goldie McCullough. It was a very efficient engine. There was a steam condenser that got the maximum power out of the steam, so the Goldie McCullough would run very evenly. Consequently, if everything was operating properly less steam was required to operate efficiently.

That winter the steam plant supplied all of Hedley's local requirements electrically, as well as supplying Mascot Mines if their diesel plant was down. We did have a hard time then supplying enough power. The engineers in the powerhouse monitored the electrical output with gauges and recording equipment. A lot of the power we produced went to a 400 hp synchronized motor which ran air compressors that supplied air for the Nickel Plate Mine, piped right up to the mine. Later on after West Kootenay put in a proper substation in Hedley and a power line up to the mine, the air compressors were run from up there.

The turbine on the river was about three miles down river from where the dam was. There was a flume that supplied the water to the head of the penstock. Even though we had screens in front of the water from the Similkameen entering this flume, a certain amount of wood would get into it. One time when the water was shut off and the turbine was shut down, I went down from the top and into the turbine and picked out all the wood and put it in a basket and hauled it out of there, because it would get jammed in the blades. When I was working in the powerhouse, the dam had not yet been re-activated. It was later on, when the mill was in operation that the dam was re-activated, until the time that West Kootenay Power put in their substation, which is still there now.

Originally, there was no power plant there. The equipment was still a water turbine, right in that same powerhouse and water came from Twenty Mile Creek. The traces of the flume can still be seen now, above the town of Hedley, from Twenty Mile Creek past where the Mascot concentrator was later installed, and into a water tank which stood above the old powerhouse in Hedley. That water ran all summer long. In November, or whenever it froze up, the Nickel Plate operation was shut down for the year, until the following spring.

The air for the mine was compressed at the bottom and pushed up the hill in a six inch line. Originally, before they had the power for the motor, the compressors were run by hydraulic turbines. That turbine was still in place down below the floor, though it was no longer in use in my time. There was a horizontal handwheel iron about 30 inches in diameter which controlled the valve that controlled the amount of water going through. That wheel was still there, and was covered by a piece of 1/4" plywood, which was used as a table for the engineers. In 1934, when I came to Hedley, both the steam plant and the river turbine were there, although neither one had been operating for a few years.

On top of the concentrator, there was a window over a huge electric motor, that drove a shaft. This shaft was over the top of all the tanks that contained cyanide solution. Some of these tanks would be up to 38 feet in diameter and about 8 or 12 feet deep, with conical bottoms, and they were all built out of California Redwood. The shaft was horizontal. At the center of each tank was a belt drive that drove a vertical shaft. On the bottom of the shaft was a propellor very similar to a ship's propellor, which kept the solution in motion. If the solution were allowed to settle, all of the ore would collect in a solid cake at the bottom, with the liquid on top, and you couldn't do anything with it.

The primary crusher was a jaw crusher. From the jaw crusher, the ore went over a short screen and onto a conveyor belt. The conveyor belt was a short belt, the discharge pulley was higher than the intake pulley, and the discharge pulley was electromagnified so that any steel

Kelowna Exploration Co. Ltd. staff, October 22, 1938. Front row: Jim Bond, Joe Bysterbosh, Collin Campbell, Falky Anderson, Herman Advocaat, Eddy Brent, Fred Lunberg, Paddy Murry, John Wirth. Middle Row: Teddy Watson, Dicky Hambly, Tommy Willy, Carl Loomer, Bill Huth, Jack Bottaro, Cy Drake, Bob Neil, Gould Winkler, Paddy Wright, Bill Lowe, Andy Winkler, Lloyd Loomer, Jake Banman, Arthur King, Bill Beale, Frank Rainbow, Louie Oldenburg, Neil McLafferty, John Jamieson. Back Row: George Mills (Mill Superintendent), W.C. Douglass (General Superintendent), George Dwyer, Ken Gibbson, Stu Munro, Todd Harris, Tommy Church, Harry Nye, Dave McLellan, Alf Egerton, Jack Ure, Freeman Grawthers, Allen McGinnis, Leo Brown, Al Close, Jack Fraser, Charlie Brown, Bill Knowles, Mr. Deodoph, Eric Hodges, Herbert Lowe, Bill Corrigian Sr.   W.V. Ring photo

that was in that belt, drill bits or anything else, would stick to it and follow the belt, while the ore went on to the screens. The steel would drop off clear of the screens into a pile at the bottom. The crushers further down were smaller and the steel would have ruined them, literally breaking them up. From there the ore was crushed small enough that it went to the 40 stamps. These 40 stamps were set up jointly in five's. The second set of windows down below was over the stamp mill. From the stamps, it was complete circulation to the mills; some called them 'rod mills', but in this case they were originally powered by flint stone, which came in ballasts on sailing ships from Norway or Scandinavia. In my time, a man stood at this screen with special rivet faced gloves, and he picked out hard rock that hadn't been broken by the crushers and threw it to one side, and that was fed into these mills to do the grinding. The hard rock ground up the soft rock. Of course from there the ore went down to the cyanide tanks. Later on after the Kelowna Exploration Company operated it, they put in a set of flotation tanks, where the chemicals and oils were mixed with the ores, and it collected the metal out of the ore and floated it over the side and down a 'launderer' as it was called.

The purpose of cyanide solution was as follows. Cyanide had an affinity for gold and literally absorbed the gold out of the solutions. There was nothing to be seen in the solution with the gold in it. It was treated with zinc dust and another chemical and went through a separate process. Eventually, in my time at the mill, the principle gold metal was recovered as a gold precipitate, and before it was dried it looked like and had the texture of black Nugget shoe polish. That was treated in an electric dryer.

They never made gold bricks in my time, but before, with the old Hedley Gold Mining Company, they actually operated a refinery and poured gold bars. They were not able to refine them to where they were mint quality.

Concentrate or residue from the flotation plant was shipped in gondola cars to Tacoma, Washington. It went to The American Smelting and Refining Company, and they recovered the copper and a certain amount of cobalt, a little cadmium, a little nickel, and the gold. The Nickel Plate property paid a penalty to the American company because of the amount of arsenic, which made the gold a black color, and was a drug on the market. The gold precipitate, was recovered. When it's dried, it was sort of an orangey color, not dust, and not really coagulated either. It was put in thin black metal containers about the size of a 15 gallon drum, about 14 in. diameter and 16 in. high. The top was brazed shut. They were put in wooden boxes made in the carpentry shop. After the box was made, and before the steel container was in it, the box was closed together and a 1/4 inch deep saw cut was put into the box in all directions, from side to side over the top. After the steel container had been put in the box and the top had been nailed on, a thin iron wire was put in the saw cut all around the box. Where the saw cuts intersected, a one inch hole had been drilled to the depth of the cut. The sealing wax and the company's seal was applied at every one of the intersections of the wire. This was then shipped to New Jersey, and that is where the material was finally put through all the final treatment to pour gold bars of a mint quality.

Before that was done, it was my duty, in the six years that I was in the assay office, to thoroughly mix this in order to get a sample. This was a day's work in itself. We had shallow milk tins about two and a half inches deep and about six inches in diameter, galvanized, to hold the sample. This sample was divided into four parts: one sample was sent to the Canadian mint; one to the vaults in our local office; one to the assay office where we assayed it, and one to a public assayer in Vancouver. When we assayed it, we would take five assays from each sample, and if all five agreed then we were satisfied that we were giving a correct figure for the value of that material. If they did not agree then we had erred in some aspect and we did it over again. Our results went into our local office. The local office also received the results from Vancouver. In our local assay office, we never saw the results from Vancouver, presumably the results agreed because we never had any complaints or problems. On two or three occasions the assayer was sick, and I was responsible for doing the assays, and I'm proud of the fact that I had no complaints. The ore concentrate was shipped to Tacoma, Washington, and the precipitate went to New Jersey.

If you drive by Hedley now, you can see how they are rehandling the tailings from down at the river. In the early days it's true that they had several accidents, where real rich values got away on them and ran down into that river, but the bulk of all that tonnage that's down there went through in the years that I worked there. When I worked in that assay office, we ran double assay every day on the tailings out of the concentrator. That was a must. The concentration of the gold was .03 oz. per ton. Two hundred tons go through every day. If you multiply that by 200 tons, even so, the amount of gold that went out there today in that whole dump wouldn't be much.

In 1947 or 1948 it was part of my job to do any snowplowing around the Nickel Plate mill site, townsite, and the one and a half miles of track down to the ore bed. The caterpillar was the same gauge as the rails, which meant that you couldn't drive the machine on the inside of the inside rail because of the power poles which were next to the bank. Therefore you had to drive with one track of the machine between the rails and the other track on the dropoff side of the rail, and as the ties were stuck out eight inches, tracks were 10 inches wide, your machine was two inches out over nothing before you started.

I was the chap who had to organize Robbie Burn's Day, and for a month I was called "Mac Bottaro".

KELOWNA MINES HEDLEY LIMITED

821 HALL BUILDING

789 WEST PENDER STREET

VANCOUVER 1, B.C.

CANADA

GOLD PRODUCTION FROM NICKEL PLATE MOUNTAIN
1904 - 1955

PERIOD	OWNERSHIP	TONS	OZS. GOLD PER TON	OZS. GOLD CONTENT	GROSS VALUE
1904 - 1929	DALY REDUCTION CO.				
	HEDLEY GOLD MG. CO.	1,226,000	0.525	643,330	$13,295,357
1934 - 1955	KELOWNA MINES HEDLEY	1,975,597	0.394	777,490	$27,212,150
TOTAL - NICKEL PLATE MINE		3,201,397	0.444	1,420,710	$40,507,507
1950 - 1955	FRENCH MINE - KELOWNA	32,463	0.800	25,960	908,600
1936 - 1949	HEDLEY MASCOT	686,625	0.370	254,051	8,891,785
TOTAL - NICKEL PLATE MOUNTAIN:		3,920,685	0.434	1,700,721	$50,307,892

GOLD VALUES USED IN ABOVE TABULATION: 1904 - 1929: $20.67 PER OUNCE

1934 - 1955 $35.00 PER OUNCE

15. THE 'NEW' NICKEL PLATE ROAD

The Penticton Herald, Thurs., July 15, 1937 -

"Hedley - Marking the beginning of the completion of the new Nickel Plate - Hedley road, the two "cats" working on the construction met on Tuesday afternoon of last week, the one working down from the top and the other up from the bottom. It is expected that the surfacing of the road will be finished in about ten days' time, when it will be possible to drive a car from Hedley over the nine miles of road to the Nickel Plate and Mascot mines, instead of journeying 53 miles to make the same trip. It is expected that an official opening will take place in the near future."

In the summer of 1937 a major thoroughfare in the road network was opened. This was the road which followed part of the old Camp Rest Trail, but swung along a niche in the rock bluff, and from there on to the Nickel Plate and Mascot. The opening of this road cut down the 53 mile Green Mountain route to 9 miles. Up to this time one could ride a skip on the tramline, scramble up the Kingston trail, or drive the backdoor route up the "old" Nickel Plate Mine Road from the Green Mountain Road.

The major reason for this route not being used earlier is the almost sheer rock face that had to be drilled, blasted, and removed to create a roadway wide enough for a car or truck.

This was the "new" Nickel Plate road, which snaked up the mountain to the Nickel Plate Mine. It was opened in 1937 and reduced the trip to the mine to 9 miles, rather than 53 via the "old" Nickel Plate road, which branched off the Green Mountain Road.　　　　Joe Harris photo

Once one had navigated the winding road of the "new" Nickel Plate road there was the rock cut to contend with. Consider that in the 1930's many vehicles still had mechanical brakes. The truck belongs to Hyl Innis Trucking and Harry Allison is seen beside the rig.   Doug Cox collection

The construction was made more difficult in that the contractors had to build from the bottom. Another drawback to the construction was that rocks pushed over at the site of the blasting plummeted down the slope and landed on part of the road below. It was necessary to have a bulldozer at the bottom to remove boulders so men and equipment could travel back and forth to the blasting site. The road, when completed, was equipped with switchbacks, back-ups, corners, steep grades and a precipitous cliff.

Drivers had to be wary at all times. Once the road was built apparently very little maintenance work was completed. If the rocks that fell on the road threatened one's suspension or oilpan, you got out and removed them. If a pothole got too deep you filled it in. The road was hard on vehicles, Model A's boiled furiously on the first climb, so one changed the water at a place where the road crossed the creek. Later models seemed to do better, but none would stand up very long to the rigours of driving the road every day as many of the mine employees did.

There were mishaps and accidents but apparently no fatalities. There was one spectacular accident when a carload of revellers coming from the Nickel Plate failed to negotiate a switchback, and rolled sideways and end over end a reported fourteen times before coming to rest on the roadbed, three switchbacks, and road levels later. The occupants of the car were uninjured possible due to the fact that there was a full carload and cars were built sturdier then. There is also the possibility that one or more may have been very relaxed, thus avoiding injury!

Auto accidents were one of the hazards of travelling mountain roads particularly the Nickel Plate Road. The auto is a Hudson Terra Plane.

Dave Innis photo

The Penticton Herald, Thurs., July 15, 1937 -

"Big Dance at Nickel Plate was Enjoyed - Hedley visits 'The Top' over new road for happy gathering

Hedley - The dance at the Nickel Plate mine on Friday night, July 9, was in every way a most successful one. Held in the large dining room of the mine buildings, the large crowd enjoyed every minute of the long evening. Novelty dances, entertainment during the dance, a peppy orchestra, all helped to fill the evening. William (Buddy) Taft, of Nelson, a visitor in Hedley, gave a fine exhibition of tap dancing, and Don Morrison of the Nickel Plate, favored the gathering with a sword dance, but unfortunately not in costume. Tasty refreshments were served in abundance.

The Community Club is anticipating many more such dances, with the erection of the new community hall foundation which is already under way.

The new Hedley-Nickel Plate road has brought the communities of Hedley and Nickel Plate closer together, and it is felt by local citizens that a great deal of thanks is due to the Hedley board of trade, the Kelowna Exploration Company, the Mascot Mine, and the provincial government, for the final completion of this much-needed link."

The Barnes family had sawmills near Princeton. Sawdust was used as a heating fuel and sold regularly to customers. Jim Barnes hauled this kind of fuel.

"I hauled sawdust off the mountain to Keremeos because I was selling sawdust. This was after dad, Frank Barnes, sold the mill out at Red Bluffs in Princeton. I figured it was up to me to keep my customers in sawdust. After trucking about three loads off Nickel Plate mountain I decided it was better for me to be unemployed than to keep trucking down that road! The old Similkameen River looks like a silver thread, and switchbacks! I used a cab over engine Maple Leaf, a truck produced by General Motors of Canada. I remember it had four wheeled brakes. You had to keep your hind wheel brakes on well ahead, so all your fronts did really was kind of slow the front down. If you had full brakes on the front, it would pull you right off the road. I was raised on what was considered the most damnedable road in British Columbia; Blakeburn to Princeton. They didn't know what they were talking about! They never went up the Nickel Plate. Hodson and Tupper had a mill up in the Nickel Plate. Walter MacIntosh was doing the skidding for them. It was just by the grace of God that I didn't become the sawyer up there. This was about 1940. The road comes round on bare rock, it's undercut, you would drop for 1400 feet straight down; no side rails then.

This is a true story, I can't remember their names, but a carload of people from Nickel Plate were going to Keremeos. They got by that dangerous bluff coming down, you would look down and there would be nothing. It was too steep for anything to grow and the rocks would roll down. They had this car and they were all drunk, and that was the thing that saved them. They had about eight of them packed in, and they went over the bank and the car turned up on her top, and she slid all the way down and hit the road down below, about 300 yards down; kinda nosed into it, four wheels in the air! Somebody came along and looked and thought they were all dead - sure as hell, there wasn't anything wrong with any of them; they were all dead drunk! They were packed in there so tight, nobody even got a scratch. That was in '40 or '41"

Lenard Wheeler was hauling logs for Art Harris with a tandem Ford 800 truck in the 1960's. "We were hauling from Nickel Plate Lake to Taylor's Sawmill. I had one log sticking out of the load a bit, and it hung up on the rock bluff just above the big drop off. That very nearly put me over the bank. There was a big log next to the drop off, and that is where the front wheel of the truck landed. After that I made sure my logs were in against the load nice and tight!"

The "new" road to the Nickel Plate Mine was upgraded during the 1990's to make the journey a little less adventuresome for the mine employees and firms bringing supplies to the mine. With the closure of the Nickel Plate Mine the road will be presumably used for accessing the Hedley Mascot Mine buildings, now being restored as a visitor's centre. The trip up and down the road might be the highlight of the trip for these visitors!

The author's vehicle on the Nickel Plate road before it was widened. The road offers a majestic view of the Similkameen valley, however, one is cautioned to stop and view rather than drive and view.

D. Cox photo

16. THE HEDLEY MASCOT MINE

By all reports, "Dour Old Dunc" was an eccentric. He had ranched in the Trout Creek district and prospected in the area, but, after his acquisition of the Mascot Fraction lived in Hedley. He had a shack right off the boardwalk, where he existed as a bachelor. He was described as a huge man, who wore an old suit with a grease stained vest, and spent much of the time sitting in his shack mumbling or talking to himself. Because of the closeness of his shack to the sidewalk, several passersby brought his one-sided conversation to his attention. "Talking to good company!" was his surly reply.

Dunc Woods' problem was that he didn't like Gomer Jones the General Superintendent, and vowed never to sell to the company, even though his obstinance caused the shut down of the mine and the mill, which was the livelihood of the two townsites. Most of Duncs' rantings concerned Gomer Jones. A person who lived near him reported seeing Mr. Woods burst out of his shack ranting and swearing, arm extended as though holding something, and kicking at this object as well. The object of his wrath was the imaginary Gomer Jones, who was being ejected from his shack.

In the year 1933, a group of young Vancouver business men, learning of the remarkable

Diamond drilling on Dunc Woods' famed eight and six-tenths acre Mascot Fraction, then owned by the Hedley Mascot Gold Mines Ltd. Note the tripod set up in the lower right hand corner.

Stocks collection photo

Hauling poles for the West Kootenay Light & Power Co. on one truck as turns were too sharp for a trailer. By the time the Mascot Mine came into operation West Kootenay Power had a sufficient electrical grid established to supply power rather than have electricity produced by water, steam, or the Similkameen River, as had been the case with the Hedley Gold Mining Co. Johnny Allison is sitting on the poles.   Dave Innis photo

discoveries by the Kelowna Company, became interested in the Mascot Fraction owned by Dunc Woods. They had associated with them an old-time prospector who was a friend of Woods, who approached him with the idea of having this group develop his property. Woods was receptive to the idea upon certain conditions. The conditions were:

1st: That this group would pledge themselves that if he entered into an agreement with them the Mascot Fraction would not be sold to any large mining company;

2nd: That the Mascot Fraction should be immediately developed, and a mill constructed for the milling of its ore;

3rd: The group would acquire certain other claims mentioned by him, including the Eagles Nest, Nick-of-Time and Copper Chief, which claims he maintained were as rich as his Fraction;

4th: That a provincial company should be incorporated to take over the holdings and that he should receive a cash payment and a certain number of shares in the company;

5th: That he should be a Director of the company when incorporated.

Mr. V.J. Creeden, General Manager of Hedley Mascot Gold Mines Ltd. in a report to shareholders dated February 28, 1946, gives us an obituary on the late Duncan Woods.

"As a human interest feature it might be of interest to you to know that Mr. Woods, who was about 80 years of age at the time the merger was affected, took a very active interest as a Director of your Company until about the time all financing had been completed and the mill nearly ready for operation, when he seemed to lose interest and suddenly seemed like an

Classifiers being hauled to the Mascot Mill which was just being built. Semi trailers were not readily available in the early 1930's, so Innis Contracting improvised by running one truck in reverse to deliver the equipment the two miles from the railroad siding to the Mascot Mine site.

Dave Innis photo

old man who had reached the end of his life. A few days after the mill first operated he peacefully passed away at the Penticton Hospital, apparently happy and satisfied that his life's ambition had been realized."

Carl Loomer was not as complimentary about the way Duncan Woods was treated by The Hedley Mascot Gold Mines Ltd.

"Dunc Woods owned the Fraction and he wouldn't sell it to the Nickel Plate Mine because he didn't like G.P. Jones. So he sold it to the Hedley Mascot instead. And they fooled him; they gave him stock, made him a shareholder on escrowed stock, they called it in those days, meaning he couldn't sell it to anybody but them. They put him in the Three Gables Hotel in Penticton and with all the years that he had lived as a bachelor and led a pretty rough life, he couldn't take this. They let him die, and not one of the company went to his funeral or burial. Wallace Knowles and L.V. Newton in Penticton buried him."

The Mascot syndicate was permitted to use the underground workings of the Kelowna Company to diamond drill the Fraction. It was discovered that there was ample ore available on the fraction itself to justify construction of a mill. There also had to be a construction of aerial tram and the mine head. The actual buildings of the Mascot Mine are located high above the town of Hedley, and are literally hung onto the rock outcrops at this point. Building the mine head on that precarious perch was an engineering feat in itself alone; tunneling and mining ore, and then sending it down to the mill was secondary.

The Hedley Mascot Mine buildings and the portal were located on the Copper Chief claim and travelled 3000 feet at the 4800 foot level, to tap the gold rich vein in the Mascot Fraction.

Riding the aerial tram to the Mascot Mine. If you ever thought your trip to work in the morning was a bit humdrum, try this one! Mine employees, other than the single men who lived in the bunkhouse at the camp, commuted to work daily on this 5600 ft. quad haulback aerial tram. A 3-ton skip load of ore would be nearing the ore dump at the Mascot Mill at the same time.

Herb Clark Estate photo

The Fraction triangle was surrounded on three sides by the Copper Cleft, the Morning, and the Iron Duke claims.

The holdings of the Hedley Mascot Gold Mines Ltd. consisted of 47 Crown Granted Mineral claims and Fractions, comprising an area of 1486.8 acres. All of the ore produced at this mine, however, came from the Mascot Fraction during the mine's fifteen years of operation.

Pat Wright, a guide and packer from Princeton, packed all the materials for the footings and towers, and before that the drill steel and supplies for the drill crew and the work crew. "We had a camp about halfway up, used to come up that pack trail, up to the Nickel Plate at the Kingston Mine, which was about three quarters of the way to central, then I used to come around the side hill to where the tram line is now. I packed in the cement for the footings,

This is one of the towers of the Mascot aerial tram system, which connected the Mascot mine to the Mascot Mill 2000 feet below! The towers kept the tram cars from hitting the mountain as it came over the brow of the cliff.   Dutchie Vanderlinde photo

bits of scrap iron and stuff. Then they had a camp there for the crew that were working there. I packed all the grub and supplies for that.

After that when I wasn't packing I worked around the mill helping frame the mill, old Joe Barman was there and Harry Messenger. We did all the framing with an adze and a crosscut saw; it was different. After they got that built I went to work at the Nickel Plate. When the war broke out I left the Nickel Plate."

Pat has gained renown as a guide and packer in this country. He regularly did outfit guides, hunters or guests into the Ashnola Mountains, or into the wilderness areas west and south of Princeton. He has now retired, his wife and partner, Jean, died of cancer.

One of the first projects that had to be completed before the buildings could be built was to drill and blast a niche in the mountainside, wide enough to allow trucks to be brought down near the building site. The truck load after it was deposited was lowered to the build-

This is the aerial tram nearing the Mascot crusher (190 ton per day) and flotation mill. The cables are lower on the left hand side because of the weight of the car which rode the cables on four sets of sheaves, which can be seen in the foreground.   Doug Cox collection

ing site on a small tram, which had been installed. Seemingly endless staircases were built to move around the camp. When one moved around camp you were either on a staircase, or catwalk, or a skip. The complex of buildings which were apparently adequate, could be described as "different". There are different roof styles, different roof pitches, different levels and buildings for many different uses. The plan centered around a portal and a tramline. The complex included huge electric motors that ran the tramline and a very sophisticated electrical system at that time, which controlled the drum for the aerial tram. The drum, on which the tramline worked, anchored into the mountainside. There was a dump system for the ore cars connected with a loading system for the tram cars. The control room and offices were located near the front of the cluster of buildings, and the cookhouse and bunkhouses were higher up the cliff.

A local firm, Interior Contracting, then owned by the Hatfields, did some of the roadwork on the road to the mine site. Harley and Phillip Hatfield both trucked supplies into the site. Phillip recalled he drove on one of the first Ford V-8's on this particular job. Jamieson Construction was the firm that did the contract work of constructing the buildings, Frank Jamieson was in charge.

The position of trucker on this particular job was as Phillip put it, "a soft touch". Trucks had to wait up to three hours for their turn to go down, unload, and come back up before the next truck could go down. The road down, however, made up for the wait.

Electrical power for this installation did not have to be generated locally as it had been

This is one of the two 3 1/2 ton Atlas battery powered locomotives which brought ore to the surface of the Mascot Mine. There was also a Mancha trammer used to move ore.   Joe Bromley photo

for the other mines. The West Kootenay Electrical company had established an extensive power grid throughout the area and was able to supply this mine with electrical power right from the start. The power line came down Main Street in Hedley, across the tailings pond, and up the mountainside beside the aerial tramline that lead to the Hedley Mascot Mine.

The operator of the aerial tram controlled the comings and goings of the ore cars by an electrical control box, similar to those on a street car. He did have an emergency brake lever which was beside the control box, but the tram operated with little or no trouble. In summer a canvas canopy protected the operator from the sun's glare, while he surveyed the community of Hedley 3500 feet below his lofty domain. During the winter, when cloud covered the valley so thickly it seemed you could almost walk on it, he could only survey the tops of Stemwinder and the Ashnola Mountains to the south. He could never see the Mascot Mill, the destination of his ore car. The only contact to the Mascot Mill was by telephone.

The main object of the aerial tram was to carry ore which had been brought from the mine in ore cars and dumped into the ore bin. When a tram car reached the top, it was stopped underneath a chute and the operator then "barred out" the gold bearing rock into the tram car below. While this was happening the other aerial tram car was at the Mascot Mill being unloaded. The Flotation Mill constructed up Twenty Mile Creek Canyon had a 150 ton daily capacity. This was increased to 200 tons and a cyanide plant added in 1941. The mill started May 5, 1936 and ran almost continuously until 1949.

Ralph Overton was in the Hedley area when Hedley Mascot Mines started construction

Herb Clark rolling a cigarette while watching the tram line at the Mascot. The year on the calendar is 1939. The electrical controls were similar to the controls on an electric passenger tram car. The brake levers were an emergency measure. Herb Clark Estate photo

of their reduction mill and the terminal of the aerial tram line. He recalled that all of the work was done by hand. The crew Ralph was with cleared the bush with axes, picks and shovels. When they had worked far enough back, another level of the foundation was poured. The construction required a cement that would not freeze, so the construction was in the late fall or early winter.

Ralph remembered that a diesel generator that had been used by the Ford Motor Company was brought in, however, all of the concrete was mixed by hand. Joe Heather was the carpentry foreman on the project. Ralph remembered that he was a good man to work for, however, everyone had to be moving at all times.

One construction project assigned to Ralph and a carpenter was rather unique. These men were to build an outhouse for Dunc Woods. The outdoor toilet was prefabricated and sent

This unique double exposure is probably due to the slow shutter speed of the Kodak Brownie camera, which was popular in the late 1930's. Herb Clark, at the controls of the Mascot tram, worked at the mine in the winter time. In other times, Herb and Joe Harris had teamed up to take guests by horse to the Cathedral Lakes, a business which later became the Cathedral Lakes Resort.

Herb Clark Estate photo

down to the Wood residence. All Ralph had to do was dig a hole and nail the structure together. "Old Dunc must have been 80 years old at that time," Ralph recalled. While Dunc was supervising the installation, he grumbled to the men, "I got all this money now what can I do with it?"

The aerial tram as previously mentioned was designed to carry ore. The cars had four concave wheels which rode two cables, there were no counter-sheaves holding the cars secure. It rode freely over the towers with only the running cable controlling the trams movement. People rode in these cars daily from work, although they were known to have come off the cables and crash to the rocks below. Sandy Brent, who worked and lived at the Nickel Plate

Herb Clark "barring out" of the ore chute at the Mascot Mine and filling one of the 3-ton skips which rode the aerial tram to the mill.
Herb Clark Estate photo

Mine, recalled an incident when he was showing some visitors the Mascot workings. While they were watching, Sandy noticed and heard an ore car coming over the brow of the mountain, bouncing and careening off the rocks, sparks flying every time the skip landed. He frantically searched for someone who knew where the safety switches were. The tram was stopped just short of the uppermost tower, which would have been toppled, had the fallen tram car continued. Another incident involves a small caterpillar-type tractor that was being taken to the mine on an aerial tram car. Near the top the wind apparently caught the skip and tipped the machine down a steep gully. No one bothered to retrieve anything, there was nothing usable left!

When the mine cookhouse and bunkhouse opened, water had to be trucked in from a spring near the Canty Mine. Sherman Broderick, had the job of hauling water for the first few months of the camp's operation. The spring was slow to fill the five hundred gallon tank

Perched precariously on a mountainside the Mascot buildings were a carpenter's nightmare. Tools that were dropped were often too difficult to retrieve. Staircases connected all buildings of the complex.   Doug Cox photo collection

on the truck, Sherman said he could make about three trips a day between waiting for the tank to fill at the Canty, and drain into the storage tank at the Mascot. Sherman later drove a truck which hauled concentrates from the Mascot Mill to a box car on the Great Northern siding from where the concentrate was shipped to Tacoma to be smelted.

Sherman recalled an experience that involved Herb Clark, a winter time employee at the Mascot, and himself, sometime around Christmas in the late 1930's. The mining company wanted a truck, which had been leased, returned to Penticton. There had been moderate snowfall earlier so Herb took a team of horses and Sherman drove the truck. The horses were to be used to pull the truck when the going got tough. The moderate snowfall at the mine site was a three foot snowfall near Apex Mountain, which was as far as the men had managed to travel after two days. Cliff Clark, Herb's brother, snowshoed up the Old Nickel Plate Road

The Mascot Mill buildings, 1942, as viewed from an approaching ore bucket. The truck was used around the mill as a utility vehicle. In the background is a carpenter working on the roof of a newly constructed building.   Don Bain photo

where he found the men stranded. After giving them some encouragement, he snowshoed down to Keremeos, phoned Interior Contracting in Penticton, and had them send a tractor and snowplow out to rescue the stranded pair. The horses did much better than the men during the ordeal. The men had thrown hay and blankets into the back of the truck; they at least had something to eat and a blanket to keep warm.

It was mentioned that Herb Clark was a winter time employee of the Mascot Mine. L.A. Clark, who surveyed and built the original Green Mountain Nickel Plate road, was Herb's grandfather. Gerald Clark, one of the first freighters to the Nickel Plate Mine, was his father. Gerry Clark died when Herb was in his teens. After completing his schooling in Penticton, Herb carried on the family ranching tradition by working on the Tweddle Ranch in Keremeos as a horseman and cowboy. It was shortly after this that he began to hunt and pack into the Ashnola and Cathedral Mountains.

Herb obtained a Class A license as a big game hunter and guide. With Joe Harris as a partner, they began the development of what is now the Cathedral Lakes Resort. During the summer and late autumn, Herb was a guide and hunter; in winter he was welcomed as an employee of the Mascot Mines. The forty acres in the Cathedral Mountains, which Herb Clark and Joe Harris petitioned the government for use as a horse pasture for their pack teams, was developed into the Cathedral Lakes Resort. The resort has been purchased by Okanagan College for use as a wilderness retreat and training area.

Herb Clark was very interested in history during his life. He was a member of the

These men have gathered in preparation for a meal at the dining hall. Each single man had a small but comfortable room. Heating and water pipes seen in the background were enclosed to prevent freezing.   Joe Bromley photo

Okanagan Historical Society and worked for the restoration of the Grist Mill in Keremeos. Herb had the foresight to obtain and keep historical photos and documents, and also remember the stories and reminiscences of the early pioneers of our country. It was, to a large extent, the Herb Clark Estate collection of photos, which formed the basis of this historical recollection. Herb's collection is by no means the only source of material. Many people were most generous with their family collections of photos and family histories, which they made available for the rest of us to recollect and enjoy.

Jack Moore, who worked at the Hedley Mascot Mine as mine foreman for most of the time that the mine and mill was in operation, recalled that the first time he arrived at the mine was via the Green Mountain Nickel Plate Road. At that time a road had been cut below Climax Bluff, and a level spot made in Climax Canyon on which the mine buildings were to be situated. A short tram line connected the turn around with the head of the mine. A skidway was constructed through some of the buildings, which was used to take the machinery down to the mine. This same skidway was used to bring the machinery out when the mine closed.

Dave Oxley worked at the Mascot mill, located in Twenty Mile Canyon. According to Dave, everyone went up the tram line, not giving the daily practice much thought. The tram was controlled by an electric motor and had a band brake on the eight foot drum which held the cable. The tram apparently never had a runaway accident as had the tram line used by the Nickel Plate Mine.

Fred Dixon and Les Graham in 1942, looking at an air tugger, which was used to drag ore to an ore pocket.
Don Bain photo

I had an opportunity to sit down with some former Mascot and Nickel Plate employees with a tape recorder running. Bob MacRae worked for the Mascot Mine and the Nickel Plate when the Mascot closed. Sandy Brent and Warren Allen worked as miners at the Nickel Plate Mine. Ernest Schneider built some of the Mascot Buildings and he and his wife Betty lived in Hedley. Paul Sharp worked as a miner at the Mascot before joining the Air Force. Here is their conversation.

Bob MacRae: "I began working underground at the Nickel Plate Mine in 1949. I was the first one to commute daily by broom from the mountain top of Hedley, and I rode the rails for almost two years before some twenty other miners joined me. I was timed at the speed of four and one half minutes top to bottom including walking short flat areas.

The tramway skips, drawn on rails by cable hoists, stopped running an hour before the mine shift was off, and resumed one hour earlier, so I walked one mile to the head of the tramway, winter and summer, in temperature ranges of 30 below zero to 90 degrees in the shade. In the morning I rode the empty ore skip up the mountain and, if lucky, one could ride the ore train cars to the mine, or walk. When the Mascot Mine, the one with buildings on the cliffs closed, I began work underground at Nickel Plate in 1949.

With the mine houses occupied, I did not relish staying in the bunkhouse so I commuted to Hedley daily. Wills Richards, an old timer at Nickel Plate, told me that in the early days, miners on their Sunday off rode a broom affair down the tramway to Hedley, and he showed me how to make one. It consisted of a piece of rubber belting and a piece of tin channeled to fit the rail and nailed to an old house broom. One sat on the broom on the rail and took off.

John Govich and friend on one of the many staircases at the Mascot Mine buildings, 1942. Don Bain photo

Pulling up on the handle threw body weight on the rubber to slow down. I found by wearing old rubber boots for brakes, and cutting the handle off the broom, that I could double my speed on the mile of rails.

One fellow built an elaborate device with a bicycle seat and brakes and asked me to follow him down. All went well until his machine tore apart, and he went sliding down the ties. Luckily he was only scraped and could walk to Hedley. He had measured his flange clearance on large track bolts and nuts, but had not realized some were pointed upward and that's what ripped his device to pieces.

I remember one occasion when I was sure I broke my own speed record. My good friend, fellow miner Jim Camarta, temporarily working on the tramway, had greased the cable rollers with a pail and ladle, and had wiped the excess grease off on the center rail that I rode. As my broom hit the grease spots my speed increased and my brakes were no good. Going too fast to jump, I rode it out and kicked off my smoking boots at the bottom. Jim swore that he did not grease the rail on purpose!

Pat Wright with his horse, Pepper, and a pack mule called Mule, at his ranch near Princeton. Pat was raised at Hedley and "pack trained" survey and building supplies to the Mascot minesite. Later, Pat and his wife, Jean, had a guiding business packing and trailing hunters and sightseers into the mountains of Southern B.C.   Doug Cox photo

I had some minor spills, but the nearest mishap I had was when a group of mule deer crossed in front of me and a large buck stood over the rail I was riding. I was too close to brake and lowered my head to pass under him. At the last second he jumped clear.

Company officials eventually banned the broom riding, but this made them short of more than twenty miners with no transportation. The company met with a committee and agreed to run a fourteen passenger bus on two shifts. Jim Camarta and myself were selected as drivers, and we compiled thousands of accident-free miles on the mountain road, under some very bad conditions. Snow, ice, rocks, cows, horses, and deer on the road, with numerous bad corners made it treacherous driving. There were times I wished I was riding carefree down the mountain on my broom!

There were three tram lines to the Mascot Mine. There was the main one from the mill, then you go through the tunnel and you come out overlooking another valley. There are two others there. I was operator of one of them for two years, from 1945 to 49. Myself and three other guys sank the deepest shaft in the Nickel Plate Mine, which went down to 3850 ft. level. We blasted and hand mucked it out. Warren Allen was there at the time."

Warren Allen recalled the time. "I worked with Bob MacRae, it was the last job I had at Nickel Plate. There were four of us in the shaft. I went up into that area in 1948 to 51. I lived up in town. I used to play for dances up there in the dance hall. The orchestra; Bob French, played the trumpet, my wife, Marion, played the piano, Cliff MacDonald was the leader. He's

Sandy Brent, at the Mascot Mine buildings overlooking the town of Hedley 3500 feet below. Sandy worked at the Nickel Plate Mine in the 1930's and 40's and often rode the Nickel Plate tram rails on a broom or a wooden device which fitted on the rails. Rubber gum boots acted as brakes. The company eventually instigated a bus service.   Doug Cox photo

gone now. He played the violin."

Don Bain worked at the Hedley Mascot from 1941 through 1943. He started as a diamond driller but later was asked to work in the engineer's office. It was here he met Rita Baxter who would later be his wife. Jack Moore, the Hedley Mascot Mine superintendent, told him, "I've got a job for you, take these ladies through the mine." One of the ladies was Rita Baxter, who was then working in the cookhouse at the Mascot Camp.

Bob McRae continued his discussion of working in the Nickel Plate Mine. "The temperature in the mine would be 40 to 45 degrees year round. The Upper Two Level was cold and icy all the time. You had to wear heavy underwear in there. It got colder and colder. Then some guy would get to blasting late, you have to blast last, and there you are freezing to death waiting for him. Nickle Plate had some of the hardest drilling and blasting rock in Canada."

Sandy Brent agreed. "There were some spots that we used to use 5 or 6 starters to start a hole. Sparks would just fly when you'd start it. I had one drift down on eight, and it was the dirtiest damn thing! We had drilled that round, clean out. It was all that curly rock; you'd blast and you'd only get just a little piece out of where the cut was. I think they blasted one, Joe Kelly was working in it, they blasted that seven times, drilled more holes, and finally it just got like a sieve, and it fell out then. They put a big sign up that told the time it took to do 3 1/2 feet in that drift; somebody nailed it on the "dry". The "dry" was the miner's change room.

This was the Canty Mine and Good Hope Mine ore bin which connected with the Mascot Mine buildings below via cable car. The mining companies each used the ore bin for one week to transfer ore to the Mascot. Ore from the Canty and Good Hope mines was trucked to this ore dump, then cabled down to the Mascot Mine, and then on to the Mascot Mill. The truck driver backed onto a ramp and made a turn before unloading the ore into the bin. There was a 6-inch guard rail before a 500 foot drop!   Doug Cox photo

Bob McRae countered, "There was some soft rock, such as on the 44 level. It was really high grade ore."

Sandy replied, "They had quite a few spots. There was one at eight, too. You couldn't drill it. But that was pretty rich."

Ernest Schneider talked about his time at the mines. "I was in the Mascot Mine in the winter, I was a carpenter. I built some of the buildings up at the Mascot. We didn't have any plans. You had to start everything with ropes. You couldn't even lay your tools down. I worked with old Phil Boudreau. He was the boss carpenter. The lumber came up on the tram. They brought it up from down below, to the Mascot, and they handled it around the best they could.

These are the derelict Mascot Mine buildings in the 1980's. When the mine closed the camp was abandoned. Time, weather and vandals have taken their toll on the structures. The mine buildings have since been rebuilt as a tourist attraction.   Doug Cox photo

One spot we built a set of stairs, 900 and some steps. They used to bring a lot of stuff in by way of the skid road, from Nickel Plate. At least they brought some of the heavy stuff down and put it down the high line to where the tunnel came out. They brought it up to there, took it through the tunnel, and then out to the mill. All the machinery came down that slide area. Everything came down that way when they started up. It was the only way of getting it in and out. The 3800 foot level was down the north side, down that other canyon, where the road come down, to the creek. We built a set of stairs from there up to the 4800 foot level.

We lived within two doors of where the slide came down. We lived in Hedley from 1937 to 1941. We used to go to the dances up at Nickel Plate up the new road. We had a 1935 Ford. The brakes were good, they held, you could always gear down. I can't remember a lot of accidents on that road.

Betty Schneider chimed in. "I can remember accidents! We just got up to the dance hall in Nickel Plate when along the fellows came to get stretchers. A car had gone over somewhere and that worried me. I knew we had to get back down that road." She continued her conversation about life as a miner's wife. "We'd always know if the fellows were coming down from the mountain. They couldn't come down, of course, if they were bringing ore down. But the nights that they weren't, we used to watch against the mountain and see their light coloured shirts and say, well, they're coming home tonight."

Ernie continued. "We stayed up there in the bunkhouses. Everybody stayed up there. They didn't charge nothing. It was wages, plus. We could stay up there free of charge. But then you

In mid-April of 1949 the Mascot ore body was depleted and the mill shut down. The portal was barred and the equipment removed. During the lifespan of the mine it had produced 700,000 tons of gold ore with a value of $8,000,000 with gold at $35 per oz. The dividends paid to shareholders was $1,250,000.   Doug Cox photo collection

didn't want to stay up there. You'd sooner get down. The board and room was good. You had all you wanted to eat, eggs, hotcakes, bacon, porridge. It was loaded. The first winter I was there we lived in a tent. Four of us in there. Boards up so far, then the tent. We'd get up in the morning, and put our feet on the floor, and the snow would be drifted right back on us. This was at Mascot. They had tents until the bunkhouses were built. About 1937, snow on the floor! Then we went down to the washroom, they used ordinary toilets, or one just ran over the hill.

When we worked on the bunkhouse, you could put your head out the window and see for miles straight down. You could throw a rock down and hear it bounce and bounce and bounce. I lost a tape one day, and when I got it it was all in pieces. We lost a hand saw another day, it went down like a cartwheel, as far as you could see it go. Everything was all fastened on ropes. It was the same if you put up a building. You had to drill holes, put drill steel in, and put the footings around it. That would hold it to the hill."

Betty continued the conversation. "The men would come down twice one week, then on a change weekend, which was every second weekend, they'd come down three times. Sandy's sister and I were great friends, and my sister lived there as well, Kathleen and Jim Doyle. We were all young and raising our families. There wasn't much to do, we went for walks, maybe to a show once in a while."

Ernie talked about the difficulty of sleeping when the men worked night shift. "They used to have the water hose on the roof of the house in the day time, to cool the house off so the guys could sleep. Down in town it was hot!"

The Hedley Mascot Gold Mine buildings shortly after they were abandoned in 1949. At this time all equipment and machinery had been removed. There is, however, smoke coming out of the chimney of one of the buildings.   Roffel/Budd photo

Paul Sharp worked at the Mascot Mine also. " I worked underground at Hedley Mascot for two years, 1940-41. Then I was conscripted to the Air Force. I was making $4.75 a day and my keep. That was good money then. I was single. I lived up on top in the Mascot Mine Buildings. We went up and down on the cable car. We stayed there free of charge. I was mucking, later on I was running an air scraper, and timbering in the mine.

You'd get one long weekend, then one short weekend, then the other two weeks you'd get a long weekend, so you'd be home for three days instead of just the two days. Then on Monday morning you took the aerial tram to get up on top, the ore bucket is what it was. When you went up on Monday morning, you lived up there for the week. You had your own room, your bunk, everything was there. It was just a little cubicle, and usually two people to a room. I stayed up there two weeks at a time. The married men came down during the week. I was single!

Most of the newcomers started on night shifts, the graveyard shift. You went in the mine at 11:00 and mucked out for a while. Then, when you had gotten all sweated up, you sat there and cut powder. The powder had to be cut before you tamped it into the hole. It has to be cut in order to press it in tight in the hole. You take a knife and slit the dynamite lengthwise and the fumes you got off that, gave you the biggest headache you ever had in your life! After you were sweated up, when all your pores were open, you got a grand headache from the fumes. Then they'd blow just before the next shift would come on. The next shift would start, just after the smoke was settled, then they'd go in and muck it out. The Nickel Plate would blast first, then the Mascot would blast. One could shake the other, or start the other dynamite blast off. It could bring rocks off your stope; you could loosen some rocks on top of the mine. You could tell who was blasting when there was a whistle, which blew in our part, and in the

Mine machinery and the heavy electric motors were skidded out of the Mascot mine site to a service road above, on this skidway. Construction materials were lowered to the mine on a temporary tram line.   Doug Cox photo

other. There were signals given to us by whistle."

Sandy Brent recalled that there was a raise coming up right alongside of Nickel Plate at the 1500 level. "We used to have to talk to the miners in the Mascot by hammering on the rock. Five knocks on a rock, and then you answer five back; that means we're going to blast. You could hear the rock tapping for a good 100 feet. It was all solid rock in the mine, no loose rocks, so the sound carried. You could hear guys dropping stuff in the Mascot Mine from where we were next door in the Nickel Plate Mine. You'd think it was right next door to you, but it could be through 200 to 300 feet of rock."

Paul Sharp continued. "I wasn't there when they broke through. The Nickel Plate went past their mine and stole a whole bunch of ore from the Mascot. When I worked there the diamond driller said there was enough ore there for 100 years, but it was so low grade. The wages they paid in those days were so low that they could afford to take out ore. Even with their inefficient methods at the Hedley Mill, they could still make money on it.

We thought an ounce and a half to the ton was good. We had one timber stope there, about the 3300 ft. level, and they were getting up to three ounces to the ton. It was so rich that they had to mix the ore with the low grade or they'd lose the gold. The mill was so inefficient, that putting three per cent through there, they'd lose it, so they had to mix it with the low grade ore."

In the years from 1936 to 1949 when the Hedley Mascot Gold Mines Ltd. was in production, 682,355 tons of ore were shipped and treated. The ore contained 223,032 ounces of gold, 54,882 ounces of silver and 1,919,832 pounds of copper. Production figures for the mine are from the B.C. Department of Mines, 1955.

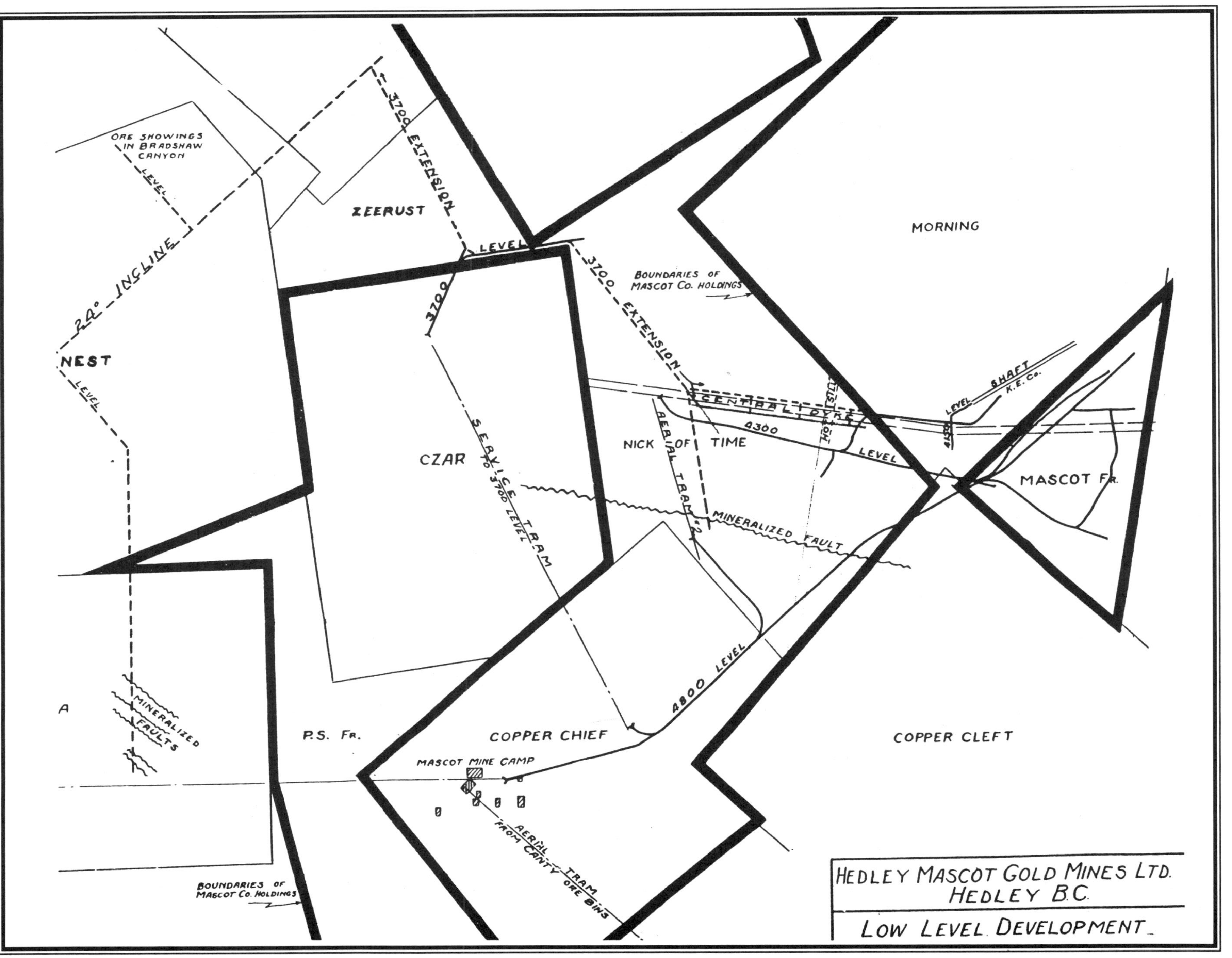

ORE SHOWINGS IN BRADSHAW CANYON
LEVEL
ZEERUST
3700 EXTENSION
24° INCLINE
NEST
LEVEL
3700
LEVEL
3700 EXTENSION
MORNING
BOUNDARIES OF MASCOT Co. HOLDINGS
SHAFT
K.E.Co.
LEVEL
CENTRAL DYKE
HORIZON
4300
LEVEL
NICK OF TIME
SERVICE TRAM
TO 3700 LEVEL
CZAR
AERIAL TRAM #2
MINERALIZED FAULT
MASCOT FR.
A
MINERALIZED FAULTS
P.S. FR.
4800 LEVEL
COPPER CHIEF
COPPER CLEFT
MASCOT MINE CAMP
AERIAL TRAM FROM CANTY ORE BINS
BOUNDARIES OF MASCOT Co. HOLDINGS
HEDLEY MASCOT GOLD MINES LTD.
HEDLEY B.C.
LOW LEVEL DEVELOPMENT.

17. OTHER MINES IN THE AREA

A mining map prepared by Frank Bailey of the mines in the area looks like clusters of stamps piled on a paper. Most of the claims are centered around areas that had shown favorable mineralization. There had been claims staked prior to those staked by Wollaston and Arundle in 1898, but there were significantly more staked after their success became known.

Many of these claims faded after a few futile years of hard work, and often a significant infusion of money. I have selected three mines in and around Hedley that were never as great producers of gold ore as the Nickel Plate and Mascot mines, but were in ways unique to the tenacity of the miners, promoters, and investors who supported them. The Golden Zone is located north of Hedley, Gold Mountain Mine is located south of Hedley, and Camp Yuniman east of Hedley.

The mining map produced by Frank Bailey, (shown on the next page) is very interesting in itself as Mr. Bailey speculates, or maps out, what the proposed development was going to be in the Similkameen Valley. Frank Bailey, mining engineer, came to the area in 1899. There was only Keremeos and Princeton in the Similkameen Valley at that time. He came from the Boundary country and purchased a tract of land near Camp Hedley from the provincial government, having been assured that the Chuchuwayha Indian Reserve and Indian Reserve No. 2 would not be thrown open for townsite purposes. The top left hand corner of the map has a hand written and signed statement from Richard H. Parkinson the Provincial Land surveyor at Fairview, B.C. stating: "I hereby certify that the Indian Reserve Boundaries as shown on this map are correct and in accordance with the Official Plans."

Mr. Bailey had R.H. Parkinson surveyed a townsite which he named Similkameen City, outside of the Indian reserves. As soon as Parkinson had surveyed Similkameen City for Bailey he started to survey Hedley City on the old bed of Twenty Mile Creek for M.K. Rodgers. Rodgers started his flume and reduction plant nearby, and by the expenditure of a considerable amount of money, and employment of an army of men, Hedley became an established fact.

Mr. Bailey, on his year 1900 map, has a proposed wagon road from his townsite that leads up the side of a cliff to the cluster of claims that would be Nickel Plate Mine. Mr. Bailey also has the area served by the Columbia and Western Railroad which follows the north side of the river, much as the right-of-way for the Vancouver-Victoria & Eastern right-of-way, and later the Great Northern, which would follow. The C. & W. however, does not cross the Similkameen River at Twenty Mile Creek.

The Columbia & Western route apparently was a serious proposal. The rail line was to turn at Keremeos Townsite, now Dick Coleman's alfalfa field, and reach Okanagan Falls via Twin Lakes. From O.K. Falls the line would travel to Penticton on the east side of Skaha Lake.

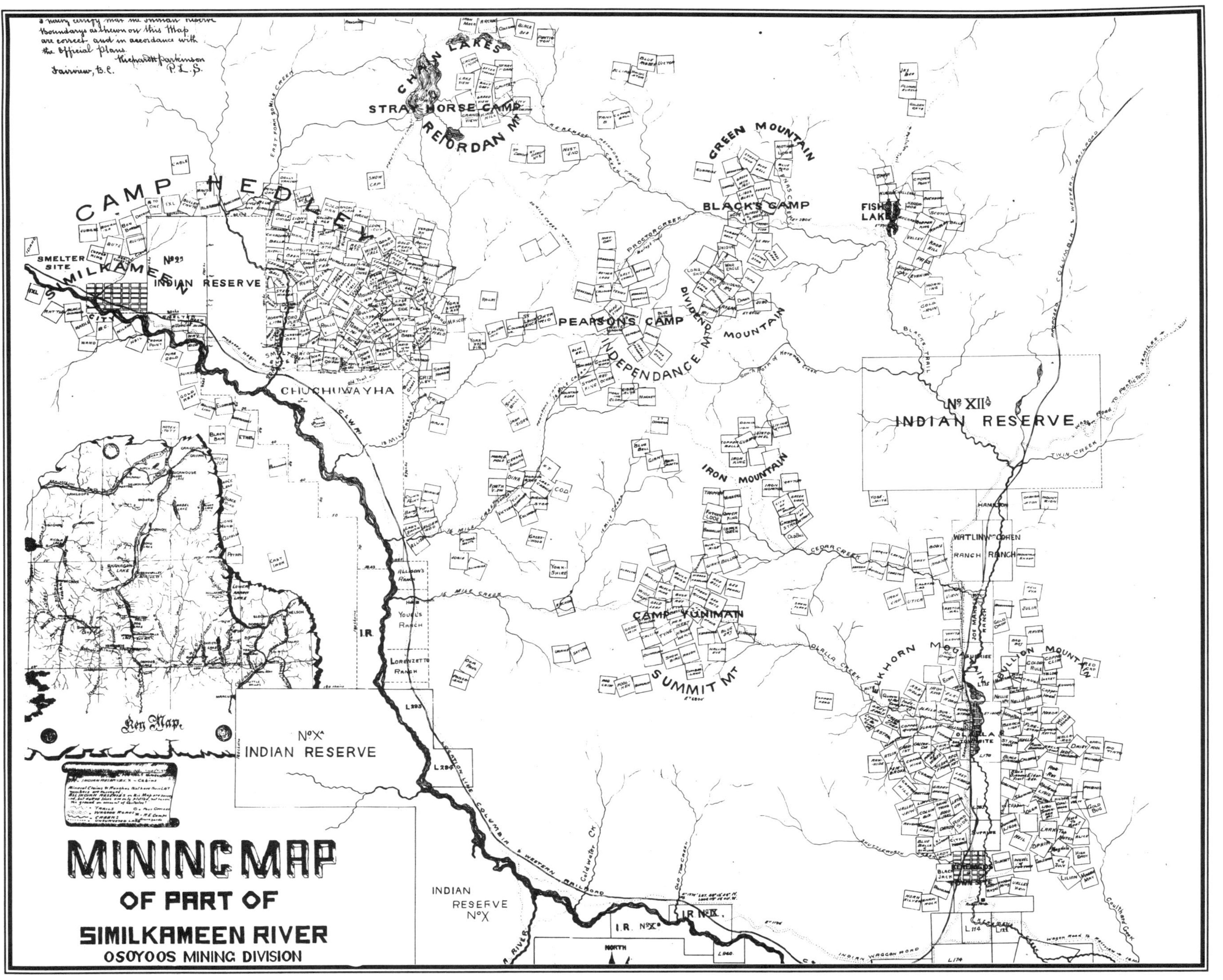

MINING MAP
OF PART OF
SIMILKAMEEN RIVER
OSOYOOS MINING DIVISION
CAMP HEDLEY
CAMP REORDAN
STRAY HORSE CAMP
CHAIN LAKES
GREEN MOUNTAIN
BLACKS CAMP
PEARSONS CAMP
INDEPENDANCE MOUNTAIN
DIVIDENCE MOUNTAIN
IRON MOUNTAIN
CAMP KININTAIN
SUMMIT MT.
FISH LAKE
No XII½ INDIAN RESERVE
CHUCHUWAYHA
SIMILKAMEEN INDIAN RESERVE No 2
SMELTER SITE
No X INDIAN RESERVE
INDIAN RESERVE No X
WATTLINN COHEN RANCH
JOE MARSELS RANCH
ELKHORN MOUNTAIN

Mr. Joe Harris, a Penticton historian, did mention that the route was to follow down much of Eastside Road and Main Street of Penticton. The reason for Front Street in Penticton being surveyed on such an angle had something to do with the right-of-way and access to the docks at Okanagan Lake for the railroad company. But that's another story!

Mr. Bailey did include a number of trails on the original 2x3 foot map, too small to see on the reduced version. The trails were usually named after the person who blazed them or used them most. The old "Camp Rest" trail through George Cahill's property is marked starting from the Chuchuwaya church. This was the first supply trail to the Nickel Plate Mine. The early ranches, which pre-date the mining era of the Similkameen Valley are marked on the map. Allison's, Youel's, and Lorenzetto's ranches are near Fifteen Mile Creek, with the Bradshaw Hotel next to the creek, which later took the name Bradshaw Creek.

GOLDEN ZONE MINE

Sherman Broderick (1908-1993) hauled supplies into the Golden Zone for a year in 1930. He apparently drove an old Chev truck with hard rubber wheels to the Golden Zone from Hedley by going up the Twenty Mile Creek Road. Sherman remembered the road as having about twenty bridges crossing the creek. After you passed the forks, there was about one bridge, after that a trestle bridge, then you climbed the sandhill. "I got stuck on that sandhill half the time. The sand would roll in and cover my tracks, and I had to get boards, sticks and everything else to fill in and make the hill."

Sherman went on to describe the road. "In the early days that's where they went. If the horse freighters wanted to freight from Hedley, they went up and over the top. That road went past the Golden Zone, on and over a little bridge, and went round to the Nickel Plate Mine.

In 1930 they had a cook house with about a dozen men in the camp. When I was up there, the machine shop caught fire. I happened to be on shift. I had to holler down the air vent of the shaft for the guys to get up there. They heard me and helped with the fire."

Sherman also used the old Chev to haul a bunch of elk out from the Naramata country out to Princeton. "Stan Robb hauled with me. We made about four trips, about four elk to the load. Finally, we had to put them in crates and haul them two at a time in the crates. We didn't make much money, about two dollars a day. They built big wing fences leading into a corral. The cowboys got up there and chased the elk down, then we loaded them onto our truck, through a chute.

The Golden Zone August, 1908. G.P. Jones with the pan. The Golden Zone consisted of four claims; the Silver Bell, Golden Zone, B.C. and Irish Boy, located in 1900, on the headwaters of Hedley Creek at an elevation of about 6000 feet. The main access to the mine was the old road, 8 1/2 miles up Twenty Mile Creek. By 1913 the property was covered with judgements from former debts contracted. As the 1913 mining report stated "Further claims and judgements were filed against the group, and another chapter added to the folly of amateurism in mining".

Pat Wright photo collection

The Golden Zone Mine made overtures to sell the property as stated in a letter dated May 28th, 1932. The following brochure plus a map accompanied the letter:

Description of Property and Assets *GOLDEN ZONE MINE, HEDLEY, B.C.*
Comprises four crown-granted claims, Viz.: Silver Bell, lot 902-S; Golden Zone, lot 903S; B.C., lot 904S, and Irish Boy, lot 905-S.
Group situated about 8 1/2 miles north of Hedley. Two roads lead to the property, one from Hedley direct, now in disuse; the other a branch from the Nickel Plate road, branching off near Nickel Plate Lake. This road leads off the Princeton-Penticton section of the southern transprovincial highway. Branch road is rough but passable. Great Northern railway and West Kootenay Company's power line run through Hedley.
Arrangements are being made with the provincial government to put the direct road from Hedley in good shape. This road is shorter and over a better grade.
Mine Equipment: *(Approximate value: $10,000)*
1 Steam boiler   1 Steam engine   1 set five stamps 1 rock crusher
1 jig table   1 steam hoist to operate to 350 ft. depth
1 No. 5 Cameron pump   1 small pump for boiler   3 hand pumps
Terms of Sale:
No. 1: Full cash value $175,000.00. Will arrange terms.
No. 2: Cast payment $100,000 (terms) Stock 100,000.
No. 3: Cast payment $100,000.00 (Payable on terms as in proposition 2).
Royalty of 50¢ per ton on all ore mine.
Signed - Yours very truly,
S.M. Nechiefman.

Princeton, B.C.
May 28, 1932

J.F. Coates, Esq.,
902 Credit Foncier Bldg.,
Vancouver, B.C.

Dear Mr. Coats:

We are enclosing herewith statement on the Golden Zone-Carrier property, and regret that we have been delayed in sending same.

Our present valuation of the property may appear high, but we are convinced that time will prove that the mine will be worth every cent of it and more.

You will appreciate that in our circumstances, development of the property is a sizeable task, and we would welcome relief. Given sufficient cash to clear our bills we will accept any reasonable terms, and will not press for full settlement until Mr. Wood's claims have been paid off. While the propositions outlined herewith approximate what we would like, if you can make, or introduce, any reasonable offer, we will be very willing to accommodate you.

Yours very truly,

S.M. Nechiefman,
SMN:T

The Gold Mountain Mine crew in 1937. In January of 1937 a 50 ton capacity mill was completed and put into operation for several months, then shut down and did not resume operations that year. Gold Mountain was south of Hedley across the Similkameen River.

Photo courtesy Reg French

Part of the mine crew at Gold Mountain Mine in 1937. Photo courtesy of Reg French

The compressor room at the Gold Mountain Mine across the Similkameen River from Hedley featured the latest in mining and electrical equipment. Although three miles from Hedley it was reached by crossing the Sterling Creek bridge between Hedley and Princeton. The calendar behind the man verifies the date as April, 1937. Photo courtesy Reg French

GOLD MOUNTAIN MINE

A short statement in the Similkameen Star Wed., January 25, 1939 - an edition which was mainly about the Hedley Rock Slide, had a short note under the heading *"Gold Mountain Mine Puts Crew to Work"*

"A crew went to work this morning at the Gold Mountain property with the intention, it is understood, of re-opening the property. It is understood the work is being undertaken by the owners and they have hopes of re-opening the mill shortly."

Nothing much more was mentioned of Gold Mountain in the mine reports or in the local papers.

George Lawrence was born and raised in the Hedley area. He now owns the ranch at Brushy Bottom, now known as "Long Acres". He recalled the place where his ranch is today:

"Brushy Bottom; when I was a kid we couldn't ride through it was so dense with brush. The cattle drives that used to go from Keremeos to Princeton would get this far and the cattle would go into the brush and they'd lose some of them. They'd stick in here all summer and when the men would come back in the fall with the cattle they'd pick them up and take them back down to Keremeos. The early traders had trouble coming through here. Some of the old stage coach roads are still visible today. They used to come through pretty well in the center of the ranch.

Horses were the only means of travelling, there weren't many cars around then. I used to pack into the mines and pack for the forestry on fires. Wherever a job came up that I could make a dollar I took it. I packed as many as ten horses. We took any kind of a horse as long as we could get a pack on and tied down. We didn't care what happened. I had one, Moby Dick, he'd wait till we got the pack on and just before we'd get it tied down, he'd turn the end of his nose up, and you knew darn well that he was going to let her go. He'd buck it off. We used to hobble, or tie one leg up on him, to pack him. We used to pack pretty solid stuff on those horses. Once you got them going they made good pack horses.

Going into Yuniman Mines with my pack train in '47 four horses had dynamite on them, and the other three had gasoline. If you got them on a switchback on a trail, I used to just turn them loose and let them follow the lead horse. The trail was such a steep trail, and switchbacks! Otherwise the horse would get halfway around the switchback, and if the one behind pulled back, you were in trouble then. Not only did we pack a lot of supplies in to the mines, but we also packed the parts in.

The heaviest part was the base of an ore car. It weighed 380 lbs. They used to make boxes for four gallon kerosene cans; two cans used to fit in these boxes. You turned two boxes upside down, one on each side of the horse, and put the ore car base on top of that. We had to swing it up in a tree and get the horse underneath and lower it down onto the pack. To rest the horse on the way up, we'd get under a tree and take some of the weight off its back. It had to be a big strong horse. Didn't hurt the horse at all, but he would be tired when he got there. About 250 pounds per horse was the usual pack needed to make it pay. We were paid two and a quarter cents per pound. I did this in 1945, and just after the war in 1946.

Johnny Knudson was the wrangler for the George Lawrence pack outfit which packed in supplies to the Yuniman Mine owned by Jack Gallagher. The square boxes on the first three pack horses are dynamite, the rounded pack is gasoline. Blasting caps? These were carried in the wrangler's shirt pocket!
George Lawrence photo

One time we took a 4,500 lb. air compressor in to the Yuniman Mine, motor and all assembled. We took the wheels off and put it on skids, and skidded it in with a big stud that we had. It was all block and tackle for seven miles. From the bottom here in the valley to the top, we had to build trails all the way through.

The stud was a grey percheron, about 1900 pounds. He was a good horse, we used him for logging as well as for breeding purposes. We got a few colts out of him. Draft horses were going out about that time. Most studs were good, dependable horses, as long as you watched them. You could turn this stallion loose with other geldings and he never bothered them. Most percherons are pretty docile.

We started on the 24th of March, skidding the compressor in. We got it in there the day before May 24th. I had guys that worked with me a lot of the time, Johnny Knudson and Tommy Smitheram. We packed hay in for the horses. Wherever we left the compressor, we tied the horse out overnight. We built a camp halfway up. When we were at the halfway mark, we worked from there.

When we used the block and tackle, we had the stud pull downhill, and he'd meet the compressor going back up. We had one triple and one double block and tackle. The compressor was going into the Yuniman claims. There was a quite a tunnel in there. It was all gold, but it never developed into much. There was quite a bit of free gold in there. The guys working in the mine used to take the free gold out of the quartz. Some of those guys came out with three or four snuff boxes of gold. They called it high grading.

Just after we got the compressor in there, they hired some miners. One of them came from Princeton. I took him in on a saddle horse on May 24th. He had been on a binge since New Year's; not sober since that time. We got him in there, and he was over in the corner bashing his head against the wall. We had the stud in there logging at the time, and one of the guys that was working with me took off about two in the morning to go for help. He ended up in the valley here about three in the morning. We tried to get a doctor to go in, but he refused to go. Finally we went back in ourselves and got him up on the old stud we were using for logging. He had gone in on a grey horse and he wouldn't get on anything except that grey horse. Of course, we never had the one he went in on, so we put him on the old grey stud. He weighed about 1900 lbs., but he thought that was fine. When we got him out, he got into town and went into the beer parlour. He got a few drinks and he was fine!

I met Barbara Pendleton in 1938, and we were married in Keremeos in 1941. We worked together on the ranch, and went into beef cattle. My dad had been in the dairying business. We started out with three head, and today it's built up to 200 head of cows, and we winter 250 head of yearlings. Barbara was a good wife, she died of cancer. We had four sons and two daughters. One of the family will be taking over the ranch when I retire."

The Gold Mountain Mine camp and crew, 1937.   Photo courtesy Reg French

For many years the Bulldog portal at the Nickel Plate Mine remained sealed. An old car frame was near the entrance and water trickled into a trough to quench the thirst of George Lawrence's cattle, which ranged here in the summer.   Doug Cox photo

This water trough was built by George Lawrence to supply water to his cattle on this part of his leased rangeland.
                                                        Doug Cox photo

Thirty years after the demise of the Kelowna Explorations' Nickel Plate Mine, the Bulldog portal was a scene of activity. A portable diesel air compressor supplied air to the men laying tram lines for the battery powered locis, which would be used to revitalize the derelict tunnels and adits.   Doug Cox photo

These men, emerging from the Bulldog Mine portal have been cleaning the tunnels of debris to provide access to the diamond drilling crews, which will follow. Some of the rotting timbers can be seen beside the track.   Doug Cox photo

Frank Holland, general manager of Mascot Gold Mines Ltd. during the exploration period, and Martin Kierands, geologist, with Sandy Brent, visitor. At this time Frank and Martin were asking Sandy, who was a Nickel Plate miner, about various aspects of the mine. Sandy's knowledge came from his experience as a miner when the Nickel Plate Mine was in operation as an underground mine.   Doug Cox photo

This was Mascot Gold Mine's offices and living quarters for the exploration crew in 1981. The original Nickel Plate Mine entrance is behind the gas tank in the center of the photo.   Doug Cox photo

Ron Simpson, geologist, looking at some drill cores left by Kelowna Exploration Co. Ltd., deep in the heart of Nickel Plate Mountain. Simpson came to the mine in 1984 when Mr. Ewanchuck became manager.   Doug Cox photo

By 1984 Mascot Gold Mines had spent over $8,000,000 on the former Nickel Plate Mine to determine the extent of the mineral reserves. These drill cores, the result of 80,000 feet of diamond drilling, are stored on the former bowling alley of the community center, started in 1937.   Doug Cox photo

This was the Nickel Plate Mine in 1981. The original Nickel Plate townsite and mine buildings had been removed, the mine entrance sealed, and the property basically abandoned when the mine closed in 1955.   Doug Cox photo

The bowling alleys and the foundation of the old community center made a convenient location for the more than 40,000 feet of diamond drill core. Mascot Gold Mines eventually completed in excess of 80,000 feet of diamond drilling to determine the extent of the mineral reserves.   Doug Cox photo

If you are a bit confused by the name changes of the companies who operated the Nickel Plate Mine, here is an outline of the companies, as outlined by Merlyn Royea, Manager of the Homestake Nickel Plate Mine:

1955 - 1964 Heirs and successors of Kelowna Mines Hedley Company

Nickel Plate Mine closed, inactive.

1964 Dundee Mines Ltd.

Conducted brief limited exploration.

1967 Giant Mascot Mines Ltd.

Optioned the property from Kelowna Mines Hedley Company, explored for copper only.

1971 Mascot Nickel Plate Mines Ltd.

Consolidated the interests of Giant Mascot Mines Ltd. (the optionee) and Kelowna Mines Hedley Co. (the owner) with the resulting ownership being 75:25. The two old antagonists finally joined forces, with the company started with Duncan Woods having the majority interest! No work done until 1978.

1980 Mascot Gold Mines Ltd.

Restructured to a publicly traded company in order to raise financing for gold exploration. Major exploration work was underway by 1984. In early 1986 ore reserve estimates justified the start of construction in June of a 1700 tonne per day open pit mine and concentrator. Production commenced in March 1987.

1988 Corona Corporation

Company was formed to amalgamate the interests of Mascot Gold Mines Ltd. and four other junior mining companies.

1991 International Corona Corporation

Restructured as a "pure gold" company, with base metal interests split off to another company.

1992 Homestake Canada Inc.

Resulted from the consolidation of International Corona Corporation into Homestake Mining Company of San Francisco. Production increased from 2700 to 4000 tonnes per day during the period 1987-1995.

1996

Ore reserves exhausted, pit completed production in July, concentrator finished processing stockpiles in October.

This air-operated mucking machine became some of the debris salvaged from the Nickel Plate Mine when Mascot Gold Mines started cleaning the adits and started some exploratory drilling. This piece of 1940's technology is now in front of the mine administration office.   Doug Cox photo

The Mascot Nickel Plate Mine in August 1985. The hill was cleared by Gary Gierlich and Jim Thompson in April and May of that year. They had to contend with various hazards, such as a steep slope, snow, and the septic tank for the Community Center where Gary did a splash down with the D-8 Caterpillar they were using for clearing.
                                                   Doug Cox photo

The foundation of the Nickel Plate Community Center built by Kelowna Exploration Co. Ltd. gives the location of the original townsite in relation to the new development. The construction of the mill site can be seen farther down the mountain.

Doug Cox photo

Heavy equipment was purchased from other mines, brought in, assembled, cleaned, and repaired in preparation for the extraction and loading of ore.   Doug Cox photo

Construction of the flotation mill at the Nickel Plate Mine site started early in 1985.
Doug Cox photo

The directors of Mascot Gold Mines at the opening of the mine, August 17, 1987.
L-R: Messers Page, Carroll, Douglas, Steen, Ewanchuck, Goodman, Leathley and Stack. Missing are Messers. Renaud, Callahay and Grafham.   Doug Cox photo

Mr. Vander Zalm, the B.C. premier at the time, was presented with a commemorative plaque and a golden shovel. At the time he was being pressured by the media and others for suggesting that welfare recipients get a shovel and go to work, hence the golden shovel. Seated are Mr. Ewanchuck, Mr. Steen, the Master of Ceremonies, Ivan Messmer MLA, and Jim Hewitt MP and Mascot Gold Mines' directors.   Doug Cox photo

Ed Lacey and his son, Howard, turned the beef on the charcoal spit to provide a succulent meal for the assembled guests.
          Doug Cox photo

Ross Innis (1911-1995), whose father Dave Innis horse freighted in the early 1900's, and later in the 1930's by truck to the Nickel Plate Mine; George Lawrence, packer and rancher; and Jim Camarta, mine employee and later a barber in Keremeos, at the mine opening. Doug Cox photo

The old graveyard, near the cattleguard on the north road access to the Nickel Plate mine, was brightened up by Mascot Mines. It was originally surrounded by a page wire fence, which had fallen down. The remains of the Chinese who were buried here were exhumed during the 1930's, taken down the tram to Hedley, and returned to China. The graveyard held others as well. Doug Cox photo

This was at the Princeton Portal in August of 1988. The Princeton Portal mainly was to find out how deep the ore beds went underneath the south pit. The mine didn't really mine any ore out of the area, it was basically for exploration. The mine could take some bulk samples out of the tunnel just to assess the ore grades prior to mining by open pit methods. It let them know there was sufficient ore to start a mine. The men are, L-R: Jim Walmsley mine manager; Peter Thomas, geologist; Nobel Larson; Paul Saxton, executive vice-president Mascot Gold Mines; and Doug Cox, photographer.   Doug Cox photo

This is a hydraulic scoop tram used for mucking out some of the materials after a blast. The scoop tram is diesel, which gives very little carbon monoxide. Motors all have large scrubbers on the exhaust systems as well, for any kind of underground engines. The main reason for the Princeton Portal was to go in on a level area and drill test holes underneath the ore body.   Doug Cox photo

These men are underneath what was to become the "south pit". The Princeton Portal gave the mine information as to the extent of the ore body prior to open pit mining. The men are L-R: Jim Walmsley; Nobel Larson; Peter Thomas and Paul Saxton.   Doug Cox photo

Doug Cox, at the last remaining building at the Canty Mine. Mr. Canty was instrumental in persuading Mr. Dunc Woods to let his 8.6 acre property be developed by the Hedley Mascot Gold Mines Ltd. The Canty mine, north of the Nickel Plate Mine, shared an ore dump with the Good Hope Mine. The ore was moved by aerial tram to the Hedley Mascot buildings and then down to the Hedley Mascot Mill. Nickel Plate Mine had started removing the overburden in the background, prior to open pit mining.   Doug Cox photo

The old tailings areas from the old Nickel Plate Stamp and Cyanide Flotation Mill are being remined by Candorado Mines at Hedley. The old stamp mill, at times, was very inefficient, and their recoveries were as low as 50 to 55%, which left a lot of gold still in the tailings. This company is reworking the old tailings area by what they call heap-leach operation. The big drag line over on the far end is putting the mine tailings into a hopper, and the tower to the right is cement. They pelletize the tailings, making the sand into round pellet sized balls. When they bring it over and put it into the heap-leach areas, they trickle cyanide solution overhead and it percolates all the way through. If they didn't pelletize it, they would lose some of their best material, and the cyanide solution would not percolate through. It would run off the sides. That's what happened at the start. The cyanide leaches the gold out. It goes right through and dissolves gold or silver, and they pick it up into these big ponds that are lined. The leachate from those heap leaching pads is picked up in the lined ponds. The solution is now called a "pregnant solution" because it contains the gold and silver in the solution and a little bit of cyanide as well.

Doug Cox photo

Yorkie's cabin is hidden in the woods near his powder shack, which was left standing by the logging crews. Yorkie was an old time rock driller who at one time ran the Nickel Plate tram line. When he retired the miners would slip him a case of dynamite for his claim, whenever he decided to seek solitude and work on his claim.   Doug Cox photo

The conveyor belts move the tailings from the riverside underneath the highway, and the belts are also used for stacking up the old tailings onto leach pads.

Doug Cox photo

19. THE DEVELOPMENTAL YEARS

During the years 1986, '87, and '88 the Nickel Plate Mine became the darling of the mining industry and the media. The free-wheeling premier, Vander Zalm, was in charge of a free enterprise government, gold had soared from $35 an ounce to astronomical prices on the London Market Exchange, and the economy was buoyant. The gold mining industry, which had experienced some doldrums was exuberant about the impending success of the Nickel Plate Mine.

At first, the media was cautious. However, they bought in when the mine started quoting vast reserves of gold ore, a mega-dollar mill, ore extraction facilities, plus jobs, where the wealth from the project would filter through the community.

The following are some of the significant headlines and opening paragraphs from Vancouver and Penticton papers:

The Vancouver Sun, Sat., Aug. 30, 1980.
Gold's new glitter stirs ghost mines at historic Hedley

HEDLEY - up on Nickel Plate Mountain, men are again searching for El Dorado, that mystical city abounding in gold, new believers converted by a glitter made even brighter now that gold has soared to more than $600 (U.S.) an ounce from $35 during Nickel Plate's heyday.

Three junior companies have each staked separate claims on Nickel Plate in the race to strike it rich, drilling in, atop of and around the 80-odd kilometres of tunnels and old workings that honeycomb the mountain towering 2,100 metres over Hedley.

The Penticton Herald, Mon., April 29, 1985.
Nickel Plate opening means more jobs

About 100 jobs will be created if plans to reopen mining operations at the old Nickel Plate property on Nickel Plate Mountain north of Hedley are finalized and the project proceeds.

The majority of the jobs in a three-shift per day operation would go to local residents says a prospectus issued by the Mascot Gold Mines Ltd. of Vancouver.

The company is proposing to set up an open pit and underground mining operation at the site, and to construct a plant with a capacity to process 1,000 tonnes of ore per day.

The Penticton Herald, Fri., Dec. 13, 1985.
$43m mine possible for Hedley

Mascot Gold Mines Ltd., says it has found sufficient gold deposits at its Hedley property to warrant a $43-million mine.

Company president Henry Ewanchuk said a production decision will be made when capital cost and financing details are completed.

If the plan goes ahead, construction will start in mid-1986 and finish in mid-1987.

The mine is projected to produce about 75,000 ounces of gold a year, which would make it the largest gold mine in B.C.

The Penticton Herald, Jan. 23, 1986.
Gold mine size, life extended

Mascot Mines will be mining from a larger open pit than originally planned when it opens its gold mine at Nickel Plate.

In a news release issued Wednesday, Mascot Gold Mines Ltd. president Hank Ewanchuk announced the redesign of the pit based on drilling since October, 1985.

He said the change will increase open pit surface reserves by 75 per cent and added the expansion will result in the overall addition of nearly three million tons to mineable reserves.

Ewanchuk said the expanded pit means the life expectancy of the mine will increase from about 6.5 years to 11 years.

The Penticton Herald, March 17, 1986.
Money for Mascot

Mascot Gold Mines Ltd. said Friday it has arranged a $50 million line of credit to develop its gold property near Hedley. The banker is the Commerce, Mascot president Henry Ewanchuk said in a release. Ewanchuk said additional funds are available if production is increased and more mill capacity is required. The mine will create about 142 jobs.

The Penticton Herald, Mon., June 9, 1986.
Mascot has nickel mine approval

Mascot Gold Mines Ltd. has been informed by the Environment and Land Use Committee that approval in principle has been granted for the Nickel Plate Mine at an expanded rate of 2,700 short tons per day.

Increases in permitting requirements needed to accommodate the increased tonnage will be handled through normal procedures.

The company had already received a revised feasibility study based on the 2,700 tons per day production rate, which indicates the project remains feasible at the new production rate, said company president Hank Ewanchuk.

A program of 20,500 feet of surface diamond drilling in 45 holes, has been recently initiated at the Nickel Plate Mine property, funded by a recent private placement of flow-through shares.

The Penticton Herald, Wed., July 16, 1986.
$70M bank deal clears way for Hedley mine

A 70-million project-financing agreement to put the Nickel Plate gold mine of Mascot Gold Mines Ltd. into production was signed Tuesday by the company and the Canadian Imperial Bank of Commerce.

The mine, near Hedley, is expected to start production in July, 1987. The initial estimated annual production will be about four million grams of gold.

"Although this is a fairly unseasoned mining company, and it's quite unusual for banks to underwrite 100 percent project financing for juniors, we went along with it because the management team had a proven track record," said Donald Worth, a Commerce vice-president.

The South Okanagan Review, July 17, 1986.
Nickel Plate set to go - $70 million financing

The new look for the old Nickel Plate open-pit gold mines, near Hedley, is planned permitted and - this week - financed to the tune of $70 million, announces its owner, Mascot Gold Mines Ltd. of Vancouver.

The start of production is projected for July 1987, and approval in principle has been given for the mine to operate at 2,700 short tons per day - up from the 1,800 short tons originally scheduled.

That means an estimated annual out-turn of 140,000 ounces of gold, for eight years, and the expected cost of production is estimated at only US$125 per ounce (compared with the worldwide average of around US$200 per ounce).

The Penticton Herald, Wed. Aug. 6, 1986.

Keremeos concerned about tailings plan

The Village of Keremeos has brought to the attention of the Ministry of Municipal Affairs its concern about a project to extract gold from the mine tailings ponds at Hedley.

Candorado Mines Ltd. proposes to extract the gold from the tailings using a heap leach technique which uses cyanide.

In a letter to the Ministry of Municipal Affairs, Mayor Robert White notes that the Keremeos Chamber of Commerce is concerned about the environmental damage that could occur as a result of the operation and particularly the effect it may have on the Similkameen River.

The Province, Sunday, Nov. 2, 1986.

Golden opportunity for Mascot Mines

Mascot Gold Mines Ltd. has come up with an unusual way to pay off a substantial part of the 70-million project financing for its Nickel Plate gold mine near Hedley.

The company has borrowed 100,000 ounces of gold for 18 months from the Canadian Imperial Bank of Commerce, at three percent annual interest. It has sold the gold at $409,80 US an ounce, from which it realized $56.9 million Cdn.

Mascot will use $40 million to pay off the first tranche of the $70 million it has borrowed from CIBC. The second tranche of $30 million can be converted to a term loan and repaid at a later date.

The Penticton Herald, Nov. 3, 1986.

More gold available new estimates show

New estimates show the Mascot gold mine near Penticton, has 6.39 million tonnes of mineable ore available.

This is an increase over original estimates, which said there was only 3.69 million tonnes. The new estimates mean the mine can now produce 2,430 tonnes per day instead of the originally projected 1,350 - 1,620 tonnes a day, and therefore increase its gold production.

Penticton Herald, Thurs., Jan. 22, 1987.

Jobs in jeopardy in mine union row

A union bid for jobs at the Mascot gold mine site near Hedley has prompted fears that the jobs of local workers are in jeopardy. In an application to the B.C. Labor Relations Board, the Building Trades Union is claiming that because of some existing collective agreements between the union and the project manager, Inter Pro, and some of the sub-contractors, only union mine construction trades are eligible to work on the site. Another union, the Christian Labor Association of Canada, is also claiming rights to control certain aspects of the job, the Herald has been told.

The Penticton Herald, Feb. 25, 1987.

Mascot Mine layoffs subject of hearings

Seven workers laid off by Mascot gold mine sub-contractor Interpro say they were terminated at the insistence of the B.C.-Yukon Territories Building Trades Council because they were non-union.

But an Interpro spokesman said today the workers were laid off because sub-contract work is winding down.

The Penticton Herald, March 11, 1987.

Royex set to sell Mascot Mines shares

TORONTO (CP) - Royex Gold Mining Corp. of Toronto announced Tuesday it is seeking to increase its control of International Corona Resources Ltd. to 51 from 40 percent in a move an analyst called "Pez-re-pellant".

Vancouver stock promoter Murray Pezim, known as The Pez, is feuding with Corona and Royex management over control of the companies. Pezim, a Corona director who built up the gold-rich company from the original Hemlo find, has launched a court action to unseat several other board members who have connections to Royex.

Royex also announced it had agreed to sell its 51-percent interest in Mascot Gold Mines Ltd. to Lacana Mining Corp, a company of which Royex holds 36 percent.

The South Okanagan Review, Aug. 6, 1987.

Golden Age

The third golden age of Nickel Plate Mine, at Hedley is just about to begin, and the most significant event in B.C.'s recent mining history - the formal re-opening of the province's largest gold mine - is now scheduled for August 17, 1987.

The Vancouver-based Mascot Gold Mines Ltd. has constructed a 2,700 short-ton per day open pit mine and gold recovery plant on site.

Reserves are estimated at 8.3 million tons with 0.14 ounces of gold per ton.

The Penticton Herald, Aug. 13, 1987.

Big gold mine reopening near Hedley on Monday

Monday will mark the most significant event in B.C.'s recent mining history: the reopening of the province's largest gold mine. Vancouver based Mascot Gold mines Ltd., has constructed a 2,700 short ton-per-day open pit mine and gold recovery plant on site. Reserves are estimated at 8.3 million tons with 0.14 ounces of gold per ton.

Formed in 1971 to exercise an option to acquire Nickel Plate, Mascot Gold Mines, led by Henry G. Ewanchuk, president, and Paul F. Saxton, vice-president of operations, continued and accelerated in 1984 the ongoing exploration and rehabilitation program that has led to the resurrection of the Nickel Plate Mine for the third time this century.

The Vancouver Sun, Aug. 15, 1987.

$70 million investment rewriting mining history

From the heart of this mountain, men took $47 million in Gold. It started in 1904, when Hedley boomed with the opening of the mill in town and the Nickel Plate Mine on the mountain top. The nearby Hedley Mascot Mine, on a claim less than an acre, mined a fortune. Finally in 1955, the great ore body of gold, silver and copper was exhausted - B.C. government historical marker overlooking Nickel Plate Mountain in the Okanagan.

HEDLEY - You know you've hit gold country when ranches carry names like Gold Dust Manor. Adventurers and prospectors, drifting north after the California gold rush, first discovered the precious metal here, midway between Princeton and Penticton, in the late 1800's. In the first half of this century, almost $50-million worth of gold was wrestled out of Nickel Plate Mountain by determined men with picks, shovels and tramlines.

The Province, Monday, Aug. 17, 1987.

Mascot Gold Mines grand opening Monday, August 17, 1987

Man's search for gold is legendary and the Grand opening of the Mascot Gold Mine on Nickel Plate Mountain is proof that the legend is alive.

Mascot Gold Mines Limited is a Vancouver based company incorporated in 1971. The company is active in mineral exploration and mine development in Canada and the western United States.

In 1971 Mascot Gold Mines acquired the Nickel Plate mineral property and began exploration in 1979.

The Nickel Plate Property is located nearly 4,000 feet above the town of Hedley in the Similkameen Valley. This historic site has remained dormant for almost 30 years. The future success of this gold mine will be a great asset to the economy of British Columbia.

The Province, Monday, Aug. 17, 1987.

Premier's stamp marks a third 'Golden Age'

When Premier Vander Zalm stamps a serial number on a bar of Nickel Plate gold later today, symbolically reopening the mine near Hedley, it will mark the most significant event in B.C.'s recent mining history.

Unworked for 30 years, the mine experienced two earlier bonanzas - 1898-1931 and 1934-1955.

Thought both times to be played out, it had yielded $50.3 million from 1.5 million oz. of gold, 190,000 oz. of silver and more than four million pounds of copper.

Now a third boom seems assured with construction by Vancouver-based Mascot Gold Mines Ltd. of a 2,700 ton-a-day open pit mine and gold recovery plant.

The Penticton Herald, Tues., Aug. 18, 1987.

Nickel Plate Mine opens a third time

"We did it. We got ourselves a mine," Mascot Gold Mines president Henry Ewanchuk told an often enthusiastic crowd of between 600 and 700 well-wishers attending the official re-opening of Nickel Plate Mine Monday afternoon.

The re-opening of the historic gold mine near Hedley, said to be the province's oldest and largest, is hoped to mark the beginning of a boost to the local economy.

Ewanchuk estimates 40 per cent of the 162 jobs created at the mine are held by those living in the Okanagan-Similkameen, mainly in Penticton. The annual payroll is expected to be more than $7 million.

The Penticton Herald, Oct. 2, 1987.

Gold mine president strikes out on his own

Hank Ewanchuk - a major figure in Canadian mining - has resigned as president of Mascot Gold Mines Ltd. and is striking out on his own to look for gold.

"I just wanted to go and run free with a small group," he said Thursday. "I wanted to build something from scratch."

It's a move dreamed of by many in the exploration departments of big mining companies.

Ewanchuk has worked for others for most of his 25-year career. His dream was made possible by financial backing from some of the biggest names in Canadian mining.

The Penticton Herald, Fri., Oct. 9, 1987.

Gold recovery firm seeks German funds

KEREMEOS - A company planning to recover gold from the tailings of the old Hedley mine says it is looking primarily to West German investors to raise the $5 million needed.

That information came from officials of Candorado Mines who met with Keremeos village council and local fish and game club representatives last night. A follow-up meeting will be held in Hedley tonight at 7 p.m. for the public to review the proposal.

The Penticton Herald, Tues., Oct. 13, 1987.

Gold recovery plans unveiled

HEDLEY - Details of the recovery process Candorado Mines Ltd. proposes to use in reclaiming gold from the old Hedley mine tailing piles were unveiled at a public meeting Friday attended by approximately 75 persons.

Candorado Mines had a model and pictures on display, and eight company representatives mingled with the crowd to answer questions.

Candorado has entered into an agreement with the One Way Foundation to recover the gold in the tailings piles on the Foundation's property. Pending government approval, Candorado will remove the tailings from the Hedley creek bank and the Similkameen river bank, carry them through an underpass beneath the highway, and will than re-pile them on higher ground where they will be processed.

The Penticton Herald, Thurs., Nov. 19, 1987.

Mascot production exceeding forecasts

Mascot Gold Mines Ltd. says initial gold production results at its Hedley operation have exceeded tune-up period forecasts.

Paul F. Saxton, president and chief executive officer, said today that by Oct. 31 the mine has produced 31, 380 ounces of gold.

"With the start-up now complete, production of 10,000 ounces of gold is expected in November," he said.

Revenues generated by the end of October were $18.8 million.

Saxton said the plant has at times run in excess of the design capacity of 2,700 tons per day and is now averaging 2,400 tons per day. Mining costs are in line with projections, he said.

The Penticton Herald, Fri., Feb. 5, 1988.

Mascot changes costing $9.75M

Mascot Gold Mines says a number of changes and equipment additions are being made to the concentrating plant at its Nickel Plate mine in a $9,750,000 program scheduled for completion in June.

The program is designed to reduce costs significantly and increase output of the plant in excess of the original planned capacity of 2,700 tons per day. Mascot said in a release following its annual meeting in Vancouver.

The new equipment includes a 42 by 65 inch gyratory crusher, a hopper preclarifier with 1,500 gallon per minute capacity, three additional drum filters each 13.5 by 17 feet and additional effluent treatment equipment.

The Penticton Herald, Fri., April 15, 1988.

Exploration work extends Mascot ore beds

Mascot Gold Mines reported today that exploration drilling from underground at its Hedley operation has extended gold ore beds, and that its operating costs have decreased significantly.

Paul F. Saxton, president and chief executive officer, also said in a news release that Mascot has informed the Canadian Imperial Bank of Commerce that the Nickel Plate mine has met a financial completion test.

Under terms of a bank loan, one of the tests of completion for the mine was the production of 31,000 ounces of gold in a 90-day period. The mine produced 32,846 ounces during a test period from Dec. 23 and March 21, Saxton said.

The Penticton Herald, Mon., Oct., 31, 1988.

Nickel Plate tailings leak studied

The Ministry of Environment is awaiting test results on water that is said to contain cyanide, which has been seeping in the area of a gold mine west of Apex Alpine ski resort.

Steve Haggarty, assistant mill superintendent, confirms Corona Corporation's Nickel Plate gold mine tailings pond leaked a small amount of water containing less than .05 parts per million of cyanide for about two days earlier this month.

The Penticton Herald, Feb. 20, 1991.

New lease on life for Nickel Plate mine

Corona Corp. is extending the life of its Nickel Plate mining operation above Hedley.

The mine had been scheduled to close at the end of October, but manager John Lovering said the company has given approval to a proposal that will extend the operation to the middle of 1993.

The plan involves extending Nickel Plate mine's north pit, which is the main operating pit.

Lovering said the approval is good news for the mine's employees, currently at 160, and for the Penticton area.

The payroll for a work force of that size has a considerable economic impact on the city, he said.

The work force was at 173 until just prior to Christmas, and Lovering said he expects it to get back to that level "in the next while".

"I'm not sure exactly what skills we will require, but there is a good possibility we will rehire most of those people again." Lovering said the development of the north pit into a third stage will mean some additional equipment, primarily a larger mining shovel and another large drill.

He said the mine continues to be conscious about the economics of the operation in light of the price of gold, currently at the $364 level compared with $416 at this time a year ago.

Corona has been operating the mine about four years.

The Penticton Herald, Fri., June 3, 1994.

Mine tops for safety

Homestake Canada's Nickel Plate gold mine near Penticton is among 12 B.C. mines recognized for their high standards of safety.

Nickel Plate received the Edward Prior trophy for open pit mines with between 200,000 and one million annual worker-hours, marking the third straight year it has won the honor.

"Compared to other mines of the same size across Canada, Nickel Plate's record is exceptionally good," said Mines Minister Anne Edwards, who presented the trophy at a ceremony in Vancouver May 27.

Nickel Plate employs 181 people in its mining operation. Last year the mine had three lost-time accidents, said employee relations and safety co-ordinator Glenn McDonald.

The Penticton Herald, July 4, 1995.

Hedley mine wins B.C. reclamation award

HEDLEY - (Staff) - Turning huge mounds of waste rock into grassy slopes has won kudos for the owners of the Nickel Plate Mine near Hedley.

Homestake Canada Inc. has been presented with the annual provincial mine reclamation award from the Ministry of Energy, Mines and Resources.

Brenda Dixon, the mine's environmental co-ordinator, says the award was in recognition of the land reclamation efforts already underway prior to the open-pit gold mine's scheduled closure in 1996.

The Penticton Western, Fri., March 15, 1996.

Nickel Plate helps workers

While the closure this fall of the Nickel Plate mine near Hedley is very bad news for the Penticton economy, the outlook for the mine's 200 employees is not so grim.

The area will lose a local employer with an $11-million payroll. However, an Industrial Adjustment Service Committee has been set up to assist those people losing their jobs

The Penticton Herald, May 6, 1996.

Mine site to be left in its natural state

HEDLEY (Staff) - While no one is happy about the closing of Nickel Plate, most are happy with the way the company is shutting down.

Nickel Plate has launched one of the most impressive reclamation projects for any mine in Canada, or in the world. They won the B.C. Mine Reclamation Award in 1994, and as the mine's general manager Tim Janke says, "We're often cited as the way it's supposed to be done."

Nickel Plate's reclamation project will take place in four phases.

The Penticton Herald, Monday, July 22, 1996.

Healing Begins

With Nickel Plate Mine set to close in October, owners are busy returning the site to nature, a process that will cost $15-20 million.

When the last crusher is silenced and the final bulldozer hauled away, Nickel Plate Mountain will return to the forest.

The public was invited Friday and Saturday to tour by bus the Nickel Plate Mine site near Hedley to view efforts owners are making to return the area to nature.

Homestake Canada Inc., an international gold mining company, acquired the mine as part of a package deal in 1993.

Exhausted ore reserves will see work at the open-pit gold mine that opened in 1986 completely shut down in October. But about 20 employees will remain on the site to oversee restoration efforts.

Mine reclamation legislation was first enacted in 1969, and since then in B.C., no mine can proceed without first detailing how owners will restore the site when mining ends.

Reclamation is an ongoing process, explained mine officials, and before a pit is excavated, owners must remove and stockpile topsoil for future use.

The total bill for site restoration will likely range between $15-20 million, said general manager, Merlyn Royea.

The Penticton Herald, August 19, 1996.

Mine toxins no match for bacteria

Development of system to clean waste water using micro-bugs earns Penticton man recognition from mining industry.

A Penticton man has been recognized for his efforts to work bugs into the system.

The Mining Association of Canada named Barry Given as one of its 30 'New Faces of Mining' for his work in using living bacteria to detoxify mine waste water.

Senior metallurgist and environmental analyst at Homestake's Nickel Plate Mine outside Penticton, Given, 49, and his team have set up a system using bacteria to neutralize such contaminants as cyanide and sulphur.

There in the "bug palace" bacteria are bred to neutralize the toxins which are by-products of the gold-milling process.

Long used in wine and beer making, friendly bacteria are now being called into play to help clean up oil spills and fuel leaks, says Given.

The Homestake Nickel Plate Mine management and the entire Nickel Plate staff have been exceptionally generous with their time and the resources to keep me, the author, up to date with the latest progress of the mine operation. Any time I wished to take photos of the operation, I was accommodated with a radio-controlled 4-wheel drive unit and a driver, who took me over the entire area and patiently explained the integrated workings of a modern mine. Even in the early exploration stages of the mine, a phone call to the office of Mascot Mines, or Frank Holland in Vancouver, resulted in clearance to photograph and explore the old workings of the Nickel Plate and Mascot buildings overlooking Hedley, which were out of bounds to the general public.

A year prior to the mine's closing, I sat down with Glenn McDonald, who was the employee Public Relations Officer and Safety Coordinator for the Homestake Nickel Plate Mine. I asked Glenn to explain the workings of the mine from ore leaving the stock pile to a gold button being poured in terms that a non-miner could understand. Here is his explanation of how it happens:

"The smaller drills are Gardner - Denver drills. These are diesel hydraulic and are what is called a downhole hammer drill. The hammer itself follows the steel, and they are specially made for drilling extremely hard rock. The rock up at Nickel Plate is extremely hard and stable. The old rock miners used very little rock bolting or timbering in their old structures under this open pit. The rock at Nickel Plate is two measures harder on the Bernel scale than even the iron ore mines in Ontario.

These are some drills working. Those holes in the waste are about 45 feet deep and the big drills are electric drills. They are run on 4160 volts, all the motors that control the drills are electric. They drill a 9 and 7/8-inch hole. The mast itself is 54 feet high and these drills can drill 54 feet straight down, and in one pass. That is just with one steel, and they can add extra steel whilst keeping on drilling.

At this time, the drilling is very close to the original Dickson Incline, it is actually down to it. We're working right through that area now at this time of the mine's life. The drills drop drill holes which are very deep. The miners will approach those large underground caverns of the old Nickel Plate Mine from 90 to 100 feet above and then drill a very close-knit pattern in a grid of about 12 feet square, and blow a hole right down into it and then backfill that stope with lowgrade material so that it just doesn't suddenly cave in on us with the weight of these machines on top.

The Ireco truck is pumping slurry into the drill holes and the fellow with the shovel is back-filling, or stemming, on top of the explosives to contain them. One cap sets off the whole blast. What we use is actually a hollow plastic tube with a very tiny bit of PETN (stands for a very long name) explosive, and it ignites all through that tube. The tube doesn't even disintegrate, but it puts off the whole blast. This explosive we use is much stronger than dynamite. The slurry can be formulated for what we call 'blasting velocities' as well. The copper mines over in the Highland Valley just use an ordinary ammonium nitrate and fuel oil, or Anfo explosive, which is a very slow accelerating explosive, but it gives lots of heave and lots of gas is formed, creating huge blocks of rock. For the hardness of the rock at Nickel Plate, we have to use a much faster explosive to shatter the rock first before the gas heaves it up.

The 1600 P.&H. shovel has a 7 cubic yard bucket on it. It is all electrically powered. The mine has three of these, and they're specially designed for the smaller 65 ton truck that's being loaded at this time. The mine has a bigger shovel than this one, a P.&H. 1900, and that has a 12 cubic yard bucket on it. It loads the larger Caterpillar trucks, which are 85 or 90 metric ton (40 - 50 cubic yards) trucks.

The trucks haul the ore to the primary crusher, which is a gyratory crusher. It's just a large cone-shaped bell in the dump pocket. The truck is just dumping a full load on top of the crusher. The crusher itself only has an inch-and-a-half throw on it, and it wobbles around in the hole and crushes the rock, some of which is four to five feet in diameter down to six inch minus. The power is supplied by an electric 450 horsepower motor underneath. The jackhammer is being operated by Wayne Severson. He is the gyro operator in the control room. Operating the hammer breaks up some of the oversize rock that won't fit down so the crusher can get hold of the rock to crush it.

Every drill hole in the pit is surveyed. Samples of each hole are taken to the assay lab, and it is marked on a grid in the engineering office. After the blast goes off we can tell where the ore and where the waste meets, and different things about the ore grades in the pit. The samples come out finely ground up. They're right from the blast hole drill, and are very fine, like coarse sand.

The company administration in 1992. L-R: Terry Notter, Maintenance Superintendent; Glenn McDonald, Employee Relations; Don Parsons, Chief Mining Engineer; Don Ingram, Mill Superintendent; Don Christianson, Mine General Foreman; Mike Dalgleish, Controller; and Greg Lang, General Manager.

Surveying in the north pit. Dougal Stock and Glenn McDonald.

After the blast, the surveyors keep track of all the blast holes - they're all assayed. Then the ore control people, out of the engineering office go out and take those maps that are on a grid and walk over top of the blast area, outlining the areas with pink ribbon. Anything within the pink ribbon area is loaded out for ore and sent down to the gyratory crusher. There are other colored ribbons for waste and low-grade ore. The waste is taken to the areas that have been mined out. We've filled in the central pit, and we've just completed the south pit overtop of Hedley, this was the first pit that we actually mined out, which was on top of the Princeton Portal exploratory tunnel. What you see here is the south pit being mined.

Jerry Leblanc, mine foreman.

This is the coarse feed pile, that is coming by conveyor into the mill to be fed into the rod mill. At this time, the mill is running between 170 and 180 tons per hour of mill feed, or 4000 tons per day.

The rod mill showing the rods, which are 4 inch rods, 18 feet long. They have a charge in the circulating drum of about 45 tons in the mill itself. Whenever the amperage drawn on the mill motor gets down to a certain level, more rods are added to keep the amperage up, that is how they keep the tonnage in there. The discharged rods can be seen in front of the drum. They are very high carbon metal, very hard. They provide the initial crush. Those rods are picked up by lifter bars, and rotate about a third of the way up the side of the drum, and fall down by gravity. They grind the ore initially before it is sent along to the ball mills.

Steel rods ready to be introduced to the rod mill to crush the gold ore. There are four 1000 hp rod mills and one 2000 hp rod mill in the plant.

This is the control room. The whole mill operation is very automated and controlled by what we call programmable logic controllers. Units can be brought up on a screen and operators can get a picture of whether valves are open or closed, densities, pH values of solutions, tank levels, whether a pump is running or stopped; anything of that nature. Garth Spencer is one of the mill foremen.

This catwalk looks directly down from the mill control room to the drum filter area. There are fourteen of those drum filters, and their sole job is to separate the solids from the liquids. They work through a vacuum effect. The drum rotates slowly around, the vacuum sucks on the material on the bottom and it rolls around, and then when it comes to the other side, a little puff of air is added to it so it falls off in a cake. This way the solids can be separated from the solutions.

The lime slaker. The mill uses quite a bit of lime to maintain high pH values. We use a cyanide leaching process. What one doesn't want in cyanide is hydrogen cyanide gas, which is very, very difficult to control. Cyanide itself is a very strong poison. It is very hazardous, but handling it in a solid or liquid form is much easier than handling it as a gas. To ensure that there is no cyanide gas formed, we raise the pH in all of our solutions that contain cyanide from a neutral of 6.9 up to 10.5 and that way there is no danger of hydrogen cyanide gas forming. Every mill that has silver or gold production uses cyanide because it is the only product that will dissolve gold or silver and put it into solution. That's how we separate it from the ore.

Cleaning up the basement floors.

Outside of the mill are the leach tanks and the thickener tank. The thickener tank is 140 feet across. In this tank, the ground slurry from the grinding mills has the water content reduced by adding a flocculent, making the solids settle out in the bottom. There are big rakes that go around in the bottom of the tank. The solids settle down there, and are picked up by underflow pumps and are pumped to the leach tanks. The water is then circulated off and reused in the system. It is a means of thickening the solution from about 35% solids to about 55% solids. Water will dilute the cyanide solution in the tanks, so we don't want too much. The mill has the capacity of maintaining a certain level of cyanide in the tanks to ensure we get maximum leaching from the cyanide and maximum recovery out of the gold ore.

The thickener tank, 140 feet in diameter.

After this, the solution is taken to the cyanide destruction plant. Liquid sulfur dioxide, which comes from Trail, B.C., is pumped into the reaction tanks. There are large compressors, and with much agitation, the sulfur dioxide oxidizes the cyanide into cyanate, which is an insoluble solid, and can be discharged down to the tailings pond. That is a process we pay royalties for, called the Inco Cyanide Destruction Process. It was patented by Inco. Inco is one of the big mining companies back east.

Nickel Plate Mine buildings with the mill in the center.

This is the last process in the refinery. These are propane fired bullion furnaces. They are heated to 2400 degrees to melt the precipitate. The precipitate is dumped into the furnaces and it's left under full temperature for a certain length of time. The wheels on the side roll the drum over. A little cart is run underneath it, and the slag and gold are poured into it.

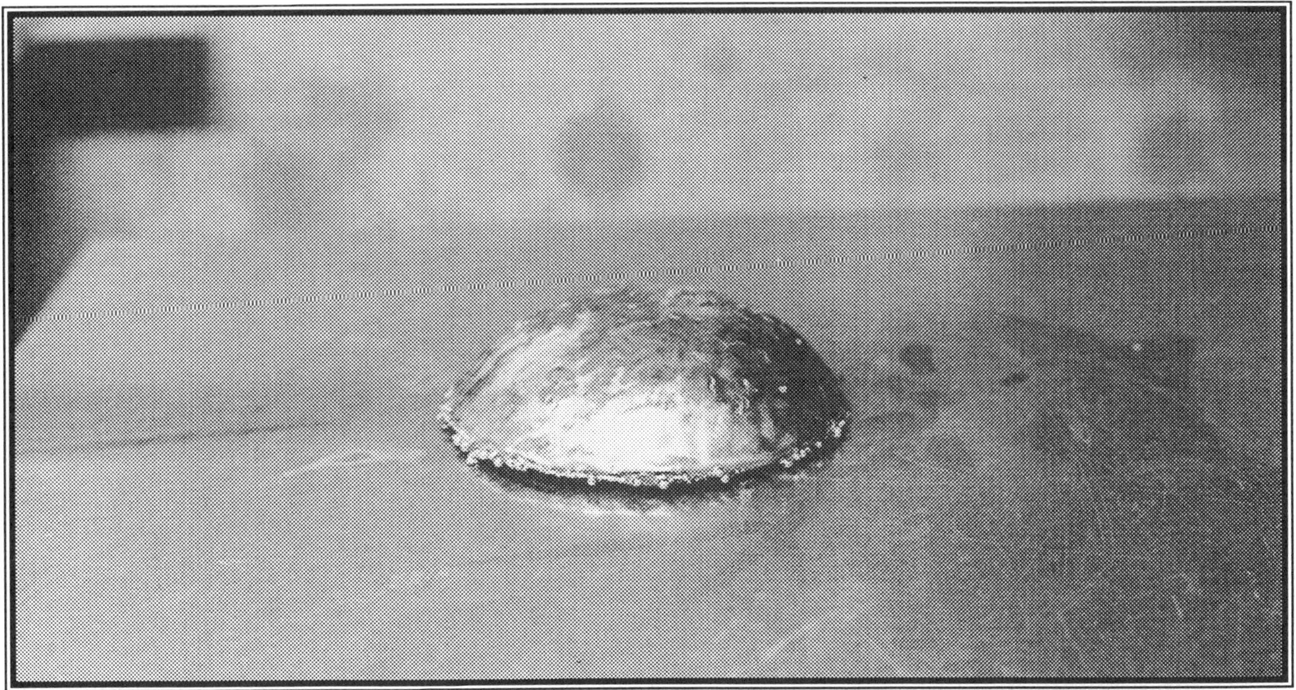

The mill actually pours what we call 'buttons' first, and the buttons are stored until there are enough to form a brick. We pour 1500 ounce doré bricks. We call them doré bricks because one can't call it gold until it is 99.99% pure. These doré bricks contain gold, silver and traces of copper. They are assayed and the mine gets paid on that. Since the mine started up, we've averaged about 85,000 ounces a year. We use a figure which we call our 'head grade' being fed into the mill, and our head grade averages about .08 ounces of gold per ton of ore.

This is discharging water into the tailings pond by what we call spigotting. There is a main line up on top of the dam, and then we spigot off the smaller lines to build a beach all the way along on the innermost side of the dam. This is called a tailings dam. This dam will have to be raised by another 12 feet, and that will be the last time in this mine's life. When we are finished, the total tailings and the water will be approximately six feet from the top. Afterwards the water will be taken out and treated, and will be discharged when it meets certain drinking water standards.

This is the floating pump-back system. The tailings have been discharged on the far end of the tailings pond, and one can see how clear the water is that comes out over here. The water is picked up by this pump-back floating barge. There are three vertical lift pumps in the pump-house, and the water is pumped back up to the mill for reuse in the system.

Just over the dam and in the trees is Cahill Creek. Just to ensure that no effluent gets out from the dam or into the creek, the mine dug all the way along on this road below the level of the Cahill Creek and put in a pumpback system that would bring the water or rainwater that would go down in there, over and into the pumpback system at that well. The Ministry of Environment gives us guidelines to meet and that is what we strive to do at all times. Homestake Mine has taken every precaution to make sure there is no leakage. When you blast, blasting agents have what you call nitrates in them, the same as you'd use in the orchards for nitrogen fertilizer. Any broken or blasted rock, if water is allowed to run through it, will generate nitrates and there's certain levels of nitrates that we have to meet for drinking water standards. Those are black bears crawling through the fence.

These pumps, are going into a pump-back well system that the mine has outside the perimeter on the downside of the dam, to make sure there are no ground waters getting away from the dam area. The dam itself is structured so that if water does get into the core of the dam, it does have what we call chimneys that will leak out through, so that it doesn't build up internal pressure and then bulge out and break. These pump-back systems are all around on the down side of the dam. These deep well pumps are about 45 feet deep and they're all automatic. They register the water flows, and the amount of the times they run. There's a backup system for every one of them. The pumps are all monitored in the control room in the mill. There are about 10 of these wells. From an environmental point of view, it is just about failure proof.

This is the mine's D10 Caterpillar doing reclamation work. The caterpillar operator is sloping down the waste dumps prior to resurfacing them with topsoil and seeding them to grass. The Caterpillar brings the dirt and small rocky material to the surface and buries the larger rocks.

This is terracing. We are looking in towards the pit area and the very top of Nickel Plate Mountain. The right side has been terraced in this way rather than coming down the slope, because it would be too steep to hold the soil. These little benches and terraces will have instant growth when seed is introduced.

The reclamation of the mine dumps is accomplished by covering them with topsoil and reseeding with grass. Protecting the topsoil from being eroded by water is completed as well. This is an example of some of the reclamation projects.

These test plots have all different types of grasses, fertilizers, depths of top soil, and every type of different condition that we could possibly think of. All these test plots are monitored on a yearly basis. We try to keep the cows out of there, but there were a lot of cattle in there last year, enjoying the good grass."

RECLAMATION

Mine reclamation legislation was first enacted in British Columbia in 1969. Prior to that date, most mining operations left it to Mother Nature to restore disturbed lands, a process that often works fairly well, but can be a long one. Since 1969 a mine cannot be developed until a satisfactory reclamation plan has been detailed, and a performance bond posted by the mining company to ensure that the work is completed.

In the last three decades, members of the public, legislators and regulators, and mining industry people, have all developed an increasing awareness that the changes imposed on the environment by human activity must be carefully weighed and controlled. We tend to think of many forms of activity as being "permanent", paving a field to make a parking lot for example, and so we don't attempt to define restoration or reclamation. Mining is, however, by definition a "temporary" activity. The ore body may be small and last only a few years, or it may be huge and last decades, but it will certainly close down sometime.

It is therefore reasonable to specify to what use mining property will be put after the economic minerals have been extracted. In a word, that is British Columbia's instruction to miners, "Restore the property to an equal or greater productive use after mining compared to that which existed before mining." In more remote areas, the end use is nearly always some combination of timber, wildlife habitat, and grazing. In more rare urban settings, the ultimate use can be residential or industrial development. In a somewhat unique case, a closed mine can be transformed into a Butchart Gardens.

At the Nickel Plate Mine, an ongoing reclamation planning has been in place since the beginning of the modern era in 1986. The program has become a major effort in the past few years, under Homestake Canada Inc. direction. Even though the profitability of the modern Nickel Plate Mine has been marginal, there was no reluctance to commit the $9 million, spent to the end of 1996, and a further $10 million projected to complete the job.

Merlyn Royea, the general manager at closure, joined Nickel Plate in mid 1996, after retiring from Cominco Ltd., where he spent over thirty years in a wide variety of engineering and management roles. He modestly admits to having, ".... no particular expertise in reclamation; we employ experts for that. I do, however, know the mining business and have a clear understanding of the required product. I'll be applying the managerial overview to get reclamation and water treatment done in a timely and efficient fashion. Rino Mihoc, the Mill superintendent, is managing the construction of the water treatment plant."

Barry Given was the senior metallurgist for the mine, and is now the environmental analyst during the reclamation period. His area of expertise is dealing with the tailings pond, which contains seventy million gallons of water, with an additional five million gallons added each month. His job is to get the cyanide and metals out of the water. To fully understand his description of the processes, one has to think back to high school and university chemistry, and biology lectures and labs, to jog our minds about chemical equations and microbiology. Barry, however, works with these concepts each day, and explained the workings in this way.

"To remove the metals and excess nutrients from the water, this mine has selected an activated sludge process. The process uses bacteria under aerated conditions to destroy thiocynate. The breakdown products from the aerated bacteria are sulphate, carbon dioxide, and ammonia. Bacteria is again used to oxidize the ammonia, which is quite toxic, and reduce it to nitrate. The nitrates cannot be introduced into the environment, so a second phase, the denitrification process is used. Here the process uses anaerobic conditions, meaning that the oxygen is removed from the water, and the bacteria is forced to get oxygen from the nitrate.

Merlyn Royea,
General Manager
of the Nickel Plate Mine.

Bary Given,
Metallurgist and
Environmental Analyst.

Brenda Dixon,
Environmental
Co-ordinator.

This breaks down the nitrate to nitrogen, which is given off as a harmless gas bubble, and the nitrate remaining is less than one part per million.

Our sister mine, a gold mine in Lead, South Dakota, with similar problems, came up with this unique microbiology process in 1970, and developed a working process by 1984. Their water chemistry is much different from ours, but we discovered a way of using the same type of bacteria but under a different process.

Metals in the water are the next concern. The process here is referred to as a high density sludge system. Ferric sulphate and lime are used to precipitate out gypsum, which is calcium sulphate. The iron also reacts with the arsenic to form a ferric arsenate, which is also precipitated out. Under the high pH readings, metals such as copper and excess iron will precipitate out. A series of thickeners recycles the biomass, and the last of the thickeners removes the bulk of the metals. The water is now ready for release into the environment. The water is being discharged into Twenty Mile Creek, which has a minimum flow of 6000 gallons per minute as opposed to Cahill Creek, which has a flow of 150 gallons per minute during the winter months. The discharged water, because it was heated at the start of the biological process, is allowed to cool before being mixed with a 6000 gallon flow."

I asked Barry how many years the reclamation crew thought it would take to treat and release the water in the tailings pond. "It will take one full year to get rid of the tailings pond water, and four or five years after that to treat the collected seepage and runoff water. That is our best estimate at this time, however, Homestake has no intention of leaving part way through the treatment process. If it takes longer to treat the water we will be here longer. We anticipate being completed by the year 2002 or 2003 at our best guess, but if it takes longer we will be here."

Brenda Dixon is the environmental co-ordinator for the mine. Her position is to co-ordinate reclamation activities such as re-sloping and re-vegetating the mine dumps. She also obtains the permits necessary to carry on any reclamation or water treatment activities at the former minesite, and is responsible for making the public aware of what is happening reclamation-wise at the mine. Her position is to liaison with the government, as she jokingly puts it, "I deal with the Ministry of Everything, Federal and Provincial, Mines and Energy, Environment and Parks, and Forestry."

Brenda was asked what the the former Nickel Plate minesite will look like when the reclamation process is completed, and the mine buildings removed. "The area will be like the surrounding area, a patchwork of open spaces and forested areas. The mine dumps have been recontoured to blend with the background, and seeded. The area will have an overlapping land use of wildlife, livestock grazing, and forestry. To achieve this the mine dumps, which were at 36 degrees, which was suitable for rock, have been re-sloped to a 26 degree slope, which is necessary to stabilize the soil spread on the slopes. The organic content of the soil is increased with the introduction of rapid growing species of grasses and legumes. The native trees and grasses, which thrive at higher elevations, are included in the reclamation program.

The area is within a registered grazing lease, which means there are cattle using the area. Goats, deer, and bear have already started to repopulate the area, even while the mine and mill were in full operation. The bears were given a respectable distance by the mine staff as they went about their duties. The food web will continue to diversify with the re-population of the area with deer, marmots, ground squirrels, and other rodents. These in turn will attract predators, such as coyotes for the smaller animals, and cougars, which prey on the larger animals. The topography will have changed, but the land use, and its productivity, will be the same."

EPILOGUE

Over the past 100 years the mineralized gold ore in Nickel Plate Mountain has been mined by various methods. The methods used are the history of hard rock mining in Southern British Columbia. When M.K. Rodgers first proved the property as a mine in 1898, the ore was mined through a hand drilled tunnel following the vein of gold ore. At one time in the early days of the mine, the gold was extracted by means of a glory hole, basically an open pit following the exposed vein of mineralization. Until 1955, with the exception of a few years when the mine was shut down, the gold ore was extracted from the bowels of the mountain through a network of stopes and nearly eighty miles of tunnels. When this operation ceased in1955, gold was priced at $35 per ounce. It is about five times as expensive to mine by underground methods, as to extract similar ore by open pit mining technology. Add to these expenses an unwieldy system of ore cars being brought to the surface by cables, transferred to a railroad and hauled to the tram line, and then reloaded into a tram car, before being taken to the concentrator mill 4000 feet below. Concentrated ore from the reduction mill was then taken by railroad down to the United States to be smelted into gold bricks.

Construction of today's Nickel Plate Mine began in May of 1986, with the first gold pour in 1987. Gold is now in excess of $350 per ounce. Using open pit mining methods, and the latest in mining technology, the mine was able to extract ore economically at 1/5 the grade of ore necessary to operate an underground mine. Electronic survey equipment can pinpoint ore locations almost instantaneously. Using this information computers can draw a three dimensional view of any ore body in about three seconds. Formerly this same activity took years to complete using layers of painted glass or balsa wood.

Milling at the Homestake Nickel Plate Mine stopped October 9, 1996. The last gold bar was poured October 14, the same year. The fixed and moveable assets will be sold. The buildings will be sold and moved. Only the former mill building housing the water treatment plant will remain, and that will only be for a few years until the reclamation process is complete. Basically the equation of .07 to .08 ounces of gold per ton of material, which could be mined at a profit, ran out. There still is mineralized gold in the mine at .02 or .03 ounces per ton, however, it is not economical to mine the ore at this time with our present technology.

In a few years most remnants of the Daly Concentrator in Hedley, and the tram line leading to the Nickel Plate Mine, will be gone. For the men who lived and worked at the mining complex, the mines will exist only in their memory. The only reminder of the forgotten mines of the eagle country will be the partially restored Hedley Mascot Mine buildings high on the cliffs above Hedley. The once boomtown and gold mining centre of Southern British Columbia is no more.

BIBLIOGRAPHY

Annual Reports of the Minister of Mines 1898-1955.
Cemeteries Division, Province of British Columbia.
The Golden Years Return, 1987. Mascot Gold Mines.
Geology and Mineral Deposits of the Princeton Map Area,
    British Columbia 1947.   H.M.A. Rice. Canada Department of Mines and Resources.
Geology and Ore Deposits of Hedley Mining District
    of British Columbia 1910. Charles Camsell. Canada Department of Mines.
Hedley Gazette 1905-1917.
Hedley Mascot Gold Mines Ltd. Report to Shareholders, Feb 28, 1947.
History of the Nickel Plate Mill and Power Plant, 1945. Harry D.Barnes.
International Corona Resources Ltd. Annual Report, 1986.
Metallurgy of Gold, Silver, Copper, Lead, and Zinc. 1908. Scranton International Textbook Co.
Ministry of Education. Province of British Columbia.
Nickel Plate Mine 1898-1932. Harry D. Barnes.
Nickel Plate Mine. May 1996. Homestake Canada Inc. Nickel Plate Mine.
Nickel Plate Mine, Summary of Process Equipment and Metallurgy. Homestake Nickel Plate Mine
Nicola, Similkameen & Tulameen Valleys. 1911. Frank Bailey, Ward Ellwood and Pound.
The Ore Deposits of Nickel Plate Mountain.
    Hedley British Columbia 1940. Paul Billingsley and C.B. Hume
Penticton Herald 1980-1997.
Preliminary Report on a part of the Similkameen District, 1907.
    Charles Camsell, Geological Survey of Canada.
Recent Cyanide Practice 1907. T.A. Richards, Mining and Scientific Press.
The Similkameen Star, January 1939.
Teacher's Record of Free Textbooks. Nickel Plate Mine School 1924-1930.
The Vancouver Sun 1939-1996.

INTERVIEWS TAPED & TRANSCRIBED

Aldredge, Edward. .Aug 1998

Allen, Warren .....June 1986

Atkinson, Arnold . . .July 1980
    Jan. 1980

Bain, Don . . . . .June 1986

Barnes, Jim Aug.1980

Biggs, Jack ....Aug. 1985

Brent, Sandy .July/Aug. 1980
   May/Sept. 1983; Sept. 1984

Brent, Edward . . . .Aug. 1987

Broderick,Sherman .Aug 1986

Bromley, Joseph . . . .Oct. 1985
    Jan 1986; Aug. 1987

Brown, Ann 1984

Clark, Herb 1979

Dickson Brenda . . .Dec. 1996

Egerton, Family . . . .Feb 1984

Estabrook, Lillian . .May 1981

French, Reg 1982

Garrison, Ernie . . . .July 1986
    March 1989

Given, Barry . . . . . .Dec. 1996

Graham, Howard 1984

Gawne, Bud . . . . . .Sept. 1981

Hardman, Myrtle 1984

Harris, Joseph . . . . .Feb 1984
    March 1986;
   July 1989; Nov 1990

Hatfield, Phillip 1982

Irwin, Arthur . . . . .May 1984

Jones, Cecil 1982

Jones, Stanley . . . . .Jan. 1997

Lacey, Edward . . . .June 1986

Lawrence, George March 1985
    May 1996

Loomer, Carl . . . . . .Oct. 1982

Liddicoat, Walter March 1984
    Jan. 1997

McDonald, Glenn . .Feb. 1996

Manuel, Randolph . .Jan. 1997

McRae, Robert . . . . .June 1986

Orser, William . . . . .June 1983

Overton, Ralph . . . .Aug. 1996

Royea, Merlyn . . . . .Dec. 1996

Schneider, Ernest . .June 1986

Schneider, Betty . . .June 1986

Sharp, Paul . . . . . . .June 1985

Smith, Mary . . . . . .June 1983

Stein, Margaret 1982

Tweddle, Alice . . . . .Jan. 1982
  Nov. 1986; Jan. 1987; Feb. 1987

Tweddle, Hal . . . . . .Dec. 1980
    Jan. 1982

VanBlaircomb, Brian May 1984

Wheeler, Dolly . . . .June 1982
    April 1996

Wheeler, Jim . . . . . .June 1982

Wheeler, Lenard . . . .Jan. 1997

Wilkinson, Bill . . . .Nov. 1996

Winkler, Andrew . . .Aug 1986

Wright, Pat . . . . .Aug 1987

Many thanks to friends, families, and public institutions of the Okanagan and Similkameen whose photos are used in this publication and are now held within the public domain.

INDEX

Brian Wilson photo

ABOUT THE AUTHOR

Doug Cox is an educator, author, and free lance historian. He and his wife, Joyce, have three children, and live on a small farm in the City of Penticton. The farm is home to a couple of pleasure horses, and they raise Dalmatian dogs and ostriches.

Doug enjoys sitting down with an old timer, with a pot of coffee or tea, and getting our exciting heritage first hand. He has published a series of history books on the Okanagan and Similkameen. He became interested in the Nickel Plate and Mascot Mines over twenty years ago, and has been collecting stories, photos, and information ever since.

There are a couple of computers in his den loaded with sophisticated word processing programs, however, Doug also has a working blacksmith shop down by the barn, just to keep in touch with reality.